New Trends in Electric Machines - Technology and Applications

Edited by Miguel Delgado-Prieto,
José A. Antonino Daviu and Roque A. Osornio Rios

Published in London, United Kingdom

New Trends in Electric Machines - Technology and Applications
http://dx.doi.org/10.5772/intechopen.100730
Edited by Miguel Delgado-Prieto, José A. Antonino Daviu and Roque A. Osornio Rios

Contributors
V. Pushparajesh, B.M. Nandish, Karlis Athanasios, Falekas Georges, Verginadis Dimosthenis, Jose A. Antonino-Daviu, Marcin Morawiec, Karolina Kudelina, Hadi Ashraf Raja, Bilal Asad, Toomas Vaimann, Juan Jose Saucedo-Dorantes, Arturo Yosimar Jaen-Cuellar, David Alejandro Elvira-Ortiz, Emmanuel Resendiz-Ochoa

First published in London, United Kingdom, 2023 by IntechOpen
IntechOpen is the global imprint of INTECHOPEN LIMITED, registered in England and Wales, registration number: 11086078, 5 Princes Gate Court, London, SW7 2QJ, United Kingdom

British Library Cataloguing-in-Publication Data
A catalogue record for this book is available from the British Library

Additional hard and PDF copies can be obtained from orders@intechopen.com

New Trends in Electric Machines - Technology and Applications
Edited by Miguel Delgado-Prieto, José A. Antonino Daviu and Roque A. Osornio Rios
p. cm.
Print ISBN 978-1-83969-851-4
Online ISBN 978-1-83969-852-1
eBook (PDF) ISBN 978-1-83969-853-8

Meet the editors

Miguel Delgado-Prieto received both his MS and Ph.D. in Electronics Engineering from the Polytechnic University of Catalonia, Barcelona, Spain, in 2007 and 2012, respectively. He is an assistant professor in the Automatic Control Department of the same university, where he carries out his teaching and research tasks. He is involved in technology transfer and project management activities as well as doctoral thesis supervision. He has co-authored more than eighty peer-reviewed scientific papers published in technical journals and conference proceedings. His research interests include industrial maintenance, fault detection algorithms, machine/deep learning, signal processing methods and embedded systems for the industry, energy, and transport sectors.

Roque A. Osornio Rios received a Ph.D. in Mechatronics from the Autonomous University of Queretaro, Mexico, in 2007. Dr. Osornio-Rios is a National Researcher level 3 with the Mexican Council of Science and Technology (CONACYT). He is currently a head professor at the University of Queretaro, Mexico. He is an advisor for more than 100 theses and co-author of more than 140 technical papers published in international journals and conferences. His fields of interest include hardware signal processing applied to mechatronics and power quality monitoring. Dr. Osornio-Rios is a fellow of the Mexican Academy of Engineering. He is also an editorial board member of the *Journal of Scientific and Industrial Research*.

José A. Antonino Daviu received both his MS and Ph.D. in Electrical Engineering from the Polytechnic University of València, Spain, in 2000 and 2006, respectively. Currently, he is a full professor in the Department of Electrical Engineering at the same university, where he develops his docent and research work. He was an invited professor at Helsinki University of Technology, Finland, in 2005 and 2007, Michigan State University, USA, in 2010, Korea University in 2014, and Université Claude Bernard Lyon 1, France in 2015. He has more than 200 publications to his credit. He is also an associate editor of multiple international scientific journals. He was an Institute of Electrical and Electronics Engineers (IEEE) Industry Applications Society (IAS) Distinguished Lecturer for 2019–2020. He received the Nagamori Award from Nagamori Foundation, in Kyoto, Japan in 2018.

Contents

Preface **IX**

Section 1
Electric Motor Maintenance **1**

Chapter 1 **3**
Perspective Chapter: Fault Detection and Predictive Maintenance of Electrical Machines
by Hadi Ashraf Raja, Karolina Kudelina, Bilal Asad and Toomas Vaimann

Chapter 2 **17**
Perspective Chapter: Data Mining in Electrical Machine Maintenance Reports
by Karlis Athanasios, Falekas Georgios, Verginadis Dimosthenis and Jose A. Antonino-Daviu

Chapter 3 **39**
Analysis of the Effects Produced by Pure Sine and Modified Sine Inverters in an Induction Motor
by Arturo Yosimar Jaen-Cuellar, David Alejandro Elvira-Ortiz, Emmanuel Resendiz-Ochoa and Juan Jose Saucedo-Dorantes

Section 2
Electric Motor Control **57**

Chapter 4 **59**
Robust Mechanism for Speed and Position Observers of Electrical Machines
by Marcin Morawiec

Chapter 5 **89**
Classical Direct Torque Control for Switched Reluctance Motor Drive
by V. Pushparajesh and B.M. Nandish

Preface

Electric machines are some of the most important technological pillars in the development of modern societies. They were first used in multiple rotatory-based applications in industry, and later found application in the energy and transport sectors. As such, the research and developments around electric machines are constantly evolving. Scientific and technological efforts in electric machines are being applied in three main areas: (1) electric motor technologies (AC/DC, synchronous/asynchronous induction, permanent magnet, reluctance, etc.), (2) supervision and maintenance strategies (fault detection, condition-based monitoring, predictive maintenance, etc.), and (3) control schemes (vector control, fault-tolerant approaches, etc.). Each field of application has its own specifications in terms of electric motor requirements and performances.

This book presents original research work on electric machine technologies and applications. It includes two sections.

The first section discusses the maintenance of electric motors, combining signal processing and artificial intelligence as a strategic methodology from which novel fault detection and identification schemes are being deployed. It also reviews predictive maintenance strategies over new industrial communication architectures, the integration of new industrial information drivers from text sources, and the effects of some renewable energy-based power schemes on electric motors.

The second section focuses on electric motor controls and their suitability to different motor technologies. It discusses direct torque control and Lyapunov-based sensorless schemes in switched reluctance, squirrel-cage induction, and permanent magnet synchronous machines as well as a doubly fed induction generator.

The chapters in this book highlight a diversity of new trends in electric machines, including electric motor technologies, supervision and maintenance strategies, and control schemes.

Miguel Delgado-Prieto, Ph.D.
Automatic Control Department,
Polytechnic University of Catalonia,
Barcelona, Spain

Roque A. Osornio Rios, Ph.D.
Faculty of Engineering,
Autonomous University of Querétaro,
San Juan del Río, Mexico

José A. Antonino Daviu, Ph.D.
Electrical Engineering Department,
Polytechnic University of València,
València, Spain

Section 1

Electric Motor Maintenance

Chapter 1

Perspective Chapter: Fault Detection and Predictive Maintenance of Electrical Machines

Hadi Ashraf Raja, Karolina Kudelina, Bilal Asad and Toomas Vaimann

Abstract

Nowadays, most domestic and industrial fields are moving toward Industry 4.0 standards and integration with information technology. To decrease shutdown costs and minimize downtime, manufacturers switch their production to predictive maintenance. Algorithms based on machine learning can be used to make predictions and detect timely potential faults in modern energy systems. For this, trained models with the usage of data analysis, cloud, and edge computing are implemented. The main challenge is the amount and quality of the data used for model training. This chapter discusses a specific version of a condition monitoring system, including maintenance approaches and machine learning algorithms and their general application issues.

Keywords: electrical machines, fault diagnostics, predictive maintenance, artificial intelligence, condition monitoring, neural networks

1. Introduction

The recent advancement in information technology, especially the integration of technology with different fields of research, has made day-to-day life convenient and opened up new research areas. One of these fields is the internet of things (IoT), which enables physical devices to communicate through the internet. The advent of these smart or intelligent devices and their implementation in industrial applications resulted in the industrial revolution, commonly known as industrial standard 4.0. These devices are not only able to communicate with each other but also able to make decisions based on defined logic or controlled remotely also referred to as cyber-physical systems. This has further paved the way for condition monitoring of electrical devices, where these devices act as data acquisition points. The collected data can then be used to monitor specific electrical machines. Further, data analysis can be done on the collected data to include fault diagnostics on these devices, including the prediction of faults [1, 2].

Industrial standard 4.0 has given way to the implementation of condition monitoring [3, 4] at a mass scale in the industry, leading toward predictive maintenance [5, 6] of electrical machines in the near future. Many companies are working on different predictive maintenance algorithms to reduce their scheduled maintenance costs. This research will further improve the effectiveness of electrical machines in the industry [7] and help to reduce unforeseen errors and faults. Most companies are also researching and finding the lifespan of the equipment based on previous patterns and external environmental variables to get the best results out of their setup. Researchers have already implemented different condition monitoring setups to maximize the potential of different electrical machines, including offshore wind turbines [8, 9], but most of this equipment is expensive and heavy.

At the moment, the industry is trying to move toward predictive and proactive maintenance to help reduce costs due to unexpected errors and faults that could have been handled before they become a more significant issue. The maintenance of electrical machines is usually divided into four phases: reactive, periodic, proactive, and periodic, as shown in **Figure 1**. Among the four phases, most of the industry is still on scheduled maintenance but is trying to move toward predictive maintenance as it is not only cost efficient but also generates a more detailed report on fault diagnostics.

With the move toward predictive maintenance, researchers are also looking for ways to utilize newer technology to get better results. The research is not only going on in this area but also in other areas like wearable devices for condition monitoring of patients to check on any abnormality [10, 11], solar-powered condition monitoring systems [12], air and noise pollution monitoring systems [13], and much more. This is because of the advancement in the technology of microcontroller boards that have given researchers more options to explore. More researchers are including these boards in their research because of their scalability. There have already been researches going on like the development of a condition monitoring system for wind turbines [14], weather sensors [15], electrical machines [16–18], autonomous vehicles [19], and robotics [20, 21]. Most of these condition monitoring systems are still in development and might need much more improvement before they can become stable and be used on a large scale. One of the most common issues is the sample rate at which data is gathered using these devices and its transmission without any data loss.

Most of the systems already in place use SCADA/PLC that are not only complex and expensive but also harder to transport [22]. One of the other issues with these systems is that although they are data acquisition points, there is no data analysis of the collected data. Hence, it is just lying there and not being utilized anywhere for fault diagnostics or being used to deduce any results. For the analysis of the collected

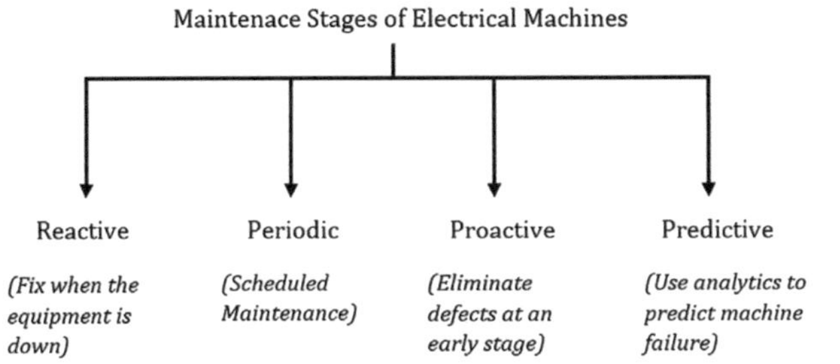

Figure 1.
Maintenance of electrical machines.

data, cloud computation is used along with edge computing, which helps analyze the data and deduce results from it. For the analysis part, machine learning algorithms are mainly used to train models based on collected data from these machines. These trained models are implemented on the cloud to get near accurate classification and prediction related to incoming data from the electrical machines. These models mainly were implemented on cloud storage or isolated servers as they need high processing power and storage space. However, now things are moving toward edge computation from the cloud. This will result in these models being implemented at the edge node where the data is being collected rather than on the cloud, which will help identify errors on the edge and further reduce the time needed to make a decision. This will also result in reduced bandwidth needed to transfer the data over the network.

This chapter discusses a concise overview of a condition monitoring system using microcontroller cards, following a small data pre-processing and analysis. Further, some light is shown on the machine learning algorithms and the training of data sets for different faults, and a short detail related to predictive maintenance is given, how it can help, and at what stage it is currently at, followed by a short conclusion.

2. Condition monitoring system

This section will discuss a particular approach to condition monitoring systems based on microcontroller boards and cloud resources. The condition monitoring system technically consists of three parts: the data acquisition system, the edge node, and the cloud. Usually, the researchers do not consider an edge node system. However, it is always better to have a local backup, computation power, and space to run some analysis if needed. The data acquisition part will consist of the microcontroller board, with the edge node being the one that helps in case of any data loss over the network.

The data acquisition part will gather data from the electrical machine using sensors. The incoming data is calibrated before transferring it through the microcontroller board to the edge node. In most cases, as the industry uses analog sensors, this part also acts as an analog-to-digital converter (ADC). The acquired data is then transmitted toward the second part that is, the edge node. The edge node acts as a local backup where the incoming data is stored in a MySQL database. The database is synced in real time with the database present in the cloud. Some pre-processing can also be done on edge, including digital filtering. The third part of the system, which is in the cloud, runs the frontend UI for the end user. It also runs diagnostics in the background on the latest synced data to look for faults. As the time difference between data acquisition from the electrical machine to showing the diagnostic results on the front end is not much, this system can also be referred to as a real-time condition monitoring system. **Figure 2** shows a rough flow chart of the implementation of a condition monitoring part.

The data is collected from the electrical machines using a microcontroller, an Arduino, or a teensy. One of the microcontroller cards (i.e., Arduino) is shown in **Figure 3**. The collected data is read through one of the analogs or digital pins on the microcontroller card, depending on the sensor used for data collection. If the sensor is IoT compatible, the data can be read over on the digital pin. In contrast, the general analog sensors used in the industry need to be calibrated and their output adjusted before they can be passed onto the microcontroller board. As the pins on the microcontroller boards do not allow a negative voltage or more than a specific voltage, before providing the data to the pins of the board, it is necessary to make sure that the sensor output is calibrated correctly. If by any chance, there is a negative voltage or

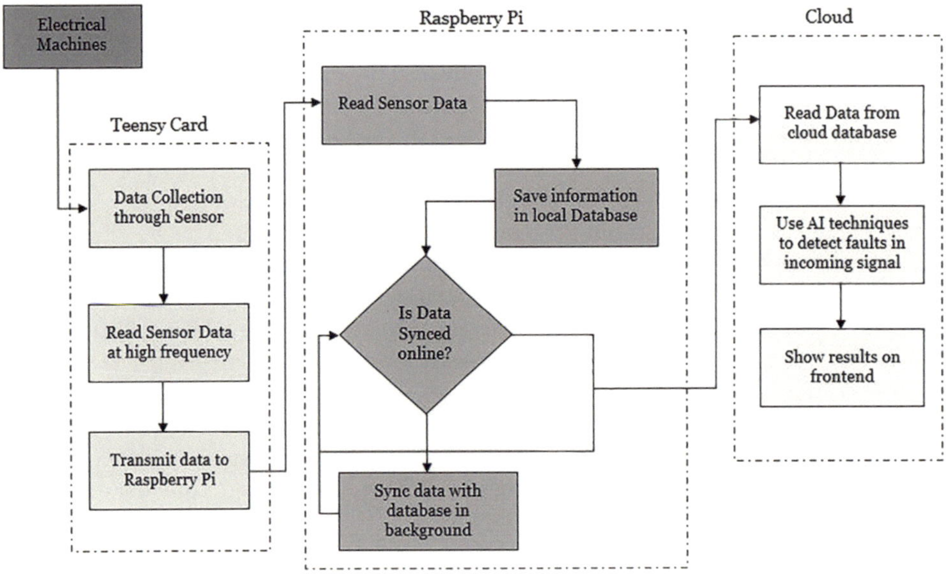

Figure 2.
Flow chart of a condition monitoring system.

Figure 3.
Arduino board.

higher voltage than the one pin can handle, there is a high chance that the board will short circuit. So, it is essential to make sure that this is handled correctly; otherwise, might end up in a short circuit of the board and with the data collected being junk without any real meaning.

Communication method	Sample rate per second
UART	1800
I2C	2600
SPI	3600

Table 1.
Comparison of sample rate for different communication methods.

Once the data is received on the board, it is then forwarded to the edge node, which is made up of Raspberry Pi. The data read here through the analog pin is at high speed. To ensure, it is transmitted at the same speed without any loss of data serial peripheral interface (SPI) connection is used between the microcontroller board and Raspberry Pi. Also, to be sure, the voltages for both the microcontroller board and Pi are different as some microcontroller boards give an output of 5 Volts at high. In contrast, Pi works with a voltage of 3.3 Volts when high, so it is also needed to ensure the transmitted values do not go over it. If a high sample rate is not needed, then UART communication should be preferred. A short description of different communication methods and their sample rate for a longer period of time is shown in **Table 1**.

The above sample rate per second is just a comparison between the speed for different communication methods for a specific microcontroller board. In this case, the microcontroller board is considered as Arduino Mega. The communication devices and other specifications, including the buffer capacity of the logger device, are the same in all three cases, that is ., Arduino Mega and Raspberry Pi. The results shown in **Table 1** are approximately the maximum sampling rates of an Arduino Mega that can be achieved when run for over a couple of days with the specific communication method without any data loss during transmission from Arduino Mega to Raspberry Pi. These specific results are hardware-dependent and changing the microcontroller board will change the speed range, example, teensy has a far better range. An experimental setup with an induction motor and analog current sensors for data acquisition is shown in **Figure 4**.

The communication method for data transmission between the microcontroller board and Pi can be decided based on the sample rate needed for the transmission. These sample rates are based on continuous data transmission from a couple of hours to days without any data loss between the transmissions. Similarly, the choice of microcontroller board might also impact the sample rate for transmission, as the newer board having better computation power gives better results. Once the data is transmitted to Pi, it is saved up in a local database and synced online simultaneously to ensure that every bit of data is synced online with the cloud without any loss. Pi also acts as a node that is capable of running analysis (like digital filtering) if needed going forward. The transmitted data is then analyzed on the cloud and based on different trained models; results are deduced whether any fault is present or not. As it is harder to understand incoming data in numerical form, the deduced results are then shown at a front end hosted on the cloud. The graphical user interface (GUI) is user-friendly and helps the end user understand the result without much information related to the system. An example of such a GUI is shown in **Figure 5**.

The GUI shown in **Figure 5** runs on the cloud with scalable resources. It mainly consists of two parts, the GUI and the diagnostic analysis running in the background. The GUI is built using PHP, whereas the diagnostic analysis primarily uses Python as the primary language, with the results saved in a MySQL database. The saved results

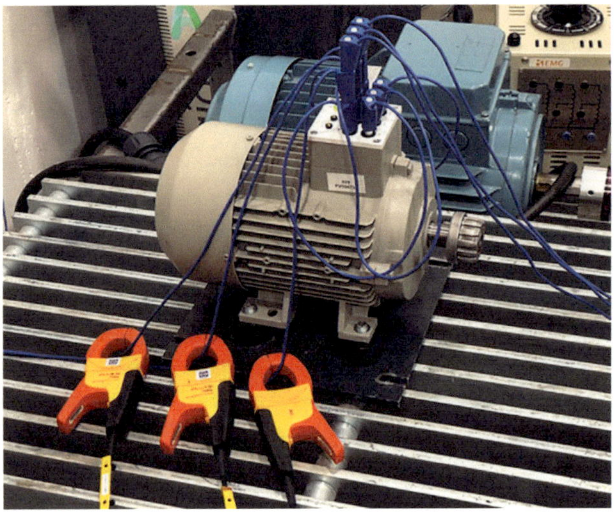

Figure 4.
Experimental setup of induction motor with analog sensors.

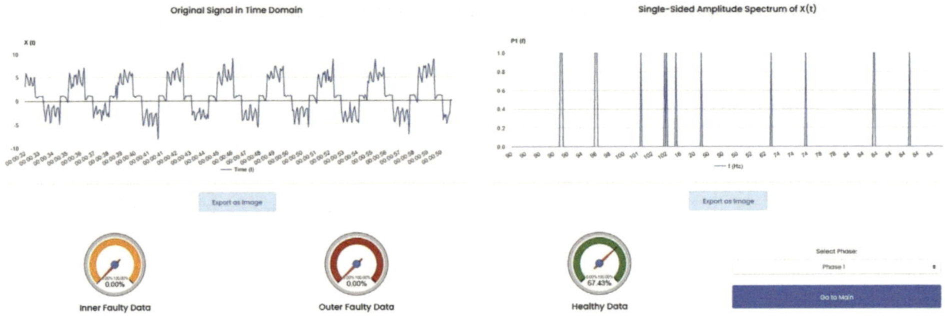

Figure 5.
Example of GUI.

are then projected on the GUI as soon as they are updated in the database. The cloud resources used here are scalable. With low data processing, that is , only one or a couple of diagnostic analyses running in the background, resources with 4 vCPU cores, 16 GB RAM, and 128 GB disk are good enough. This can be further scaled up depending on the number of diagnostic analyses and edge devices connected with the cloud, that is, increased incoming data flow.

Further, analysis results can also be shown on the GUI including the chance of a fault occurring in each phase and the option to control the electrical machine remotely if power to the machine is routed through the microcontroller board. Hence, there are multiple ways this system can be extended further. This can help the end user to understand the situation of the electrical machine in more detail. This can also help to identify which phase of the electrical machine or which part of the machine is generating issues, which can further reduce the time taken to identify the root cause of the fault. This helps maintenance teams in reducing the time needed to fix it and decide whether the fault needs to be fixed urgently or can be done later.

3. Data pre-processing and analysis

The incoming data needs to be pre-processed before it can be used for analysis. In this chapter, the analyses are focused on the steady-state operation. As the data is coming in the time domain and is raw, it is needed to make sure whether it can be utilized for the need or not. To detect faults in the early stage, it is reasonable to consider small frequency components by taking Fourier transforms of the incoming signal. For effective fault detection, different operating conditions must be considered, such as control environment, load, ambient environment, etc. **Figure 6** presents the current frequency spectra of a motor with broken rotor bars in several control modes—grid fed, scalar, and directtorque control. As seen, a significant shifting in frequency components occurs between the signals in different control modes. This is important to be considered during the model training.

At the same time, load also should be considered. **Figure 7** presents the current frequency spectra of a motor with broken rotor bars under different loads. It is seen that the behavior of the signal changes as the load increases.

In both cases, there are two regions to be studied to make predictions. Firstly, the frequency range of 0–500 Hz, where the impact of the fault is the highest on even harmonics. Specifically, the most prominent are harmonics on 50, 250, and 350 Hz. Besides, harmonics at 750 Hz can be important to be studied and considered for fault prediction.

The data is first converted into the time domain and sampled, according to the sampling frequency to make sure we have enough cycles. **Figure 8** shows an example of sample data set in the frequency domain. As the time domain does not have significant components based on which healthy and faulty data can be distinguished; hence, the data is converted into the frequency domain first and the frequency spectrum is analyzed to find the specific difference between the healthy and the faulty electrical machines.

The frequency spectrum of a faulty electrical machine includes different frequency components, usually not present in a healthy electrical machine frequency spectrum. Identifying those components and utilizing different analyzing techniques to identify them in the incoming data is part of fault detection. Including those

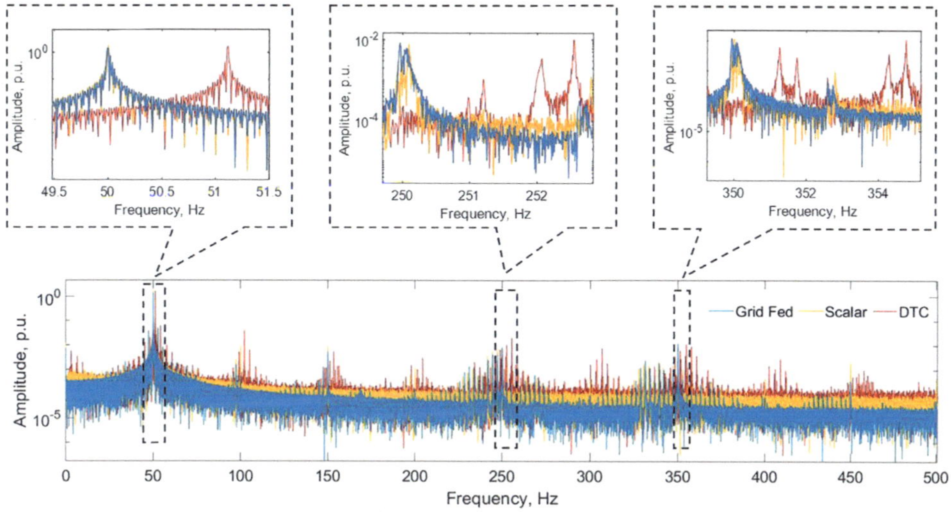

Figure 6.
Current frequency spectra of a faulty motor under different control modes.

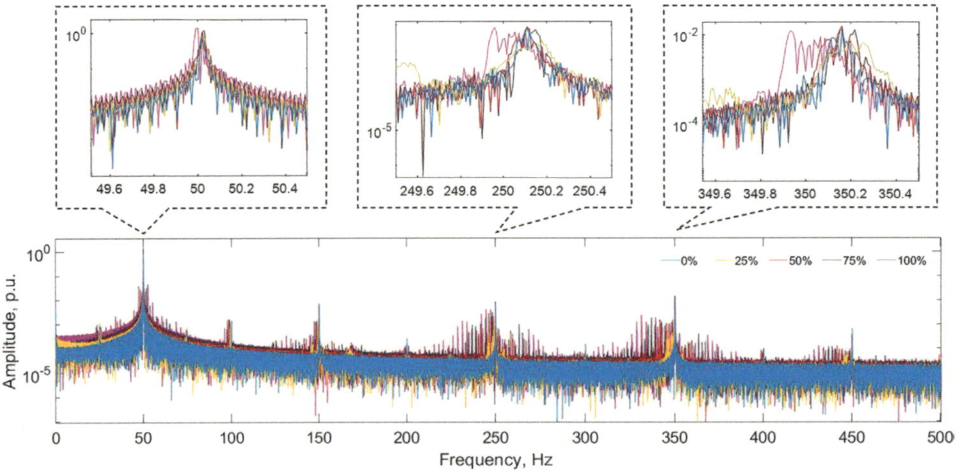

Figure 7.
Current frequency spectra of a faulty motor under different loads (grid fed).

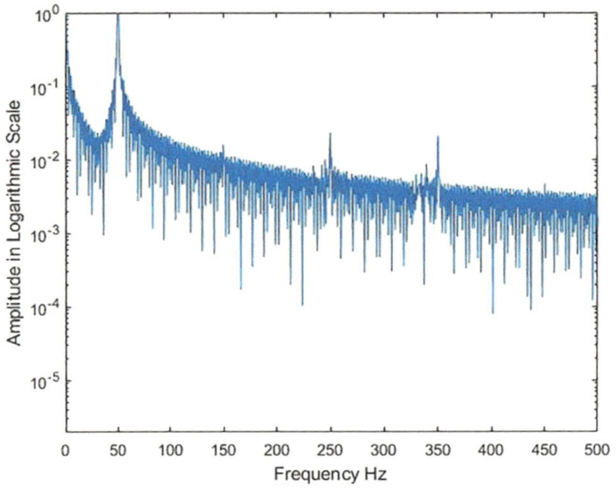

Figure 8.
The frequency spectrum of the signal to be trained.

frequency components to extract as features for training different machine learning, models can help to identify electrical machines' faults. Fault detection can be divided into two parts: signal processing and machine learning trained algorithms. Different analyses based on fast Fourier transforms can be used for the signal processing part.

4. Machine learning algorithms

The most common technique used for the detection of faults at the moment is utilizing machine learning trained algorithms. With the advent of artificial intelligence, making self-learning or systems with the aptitude for the decision has helped streamline multiple processes. Machine learning algorithms help to create a complex

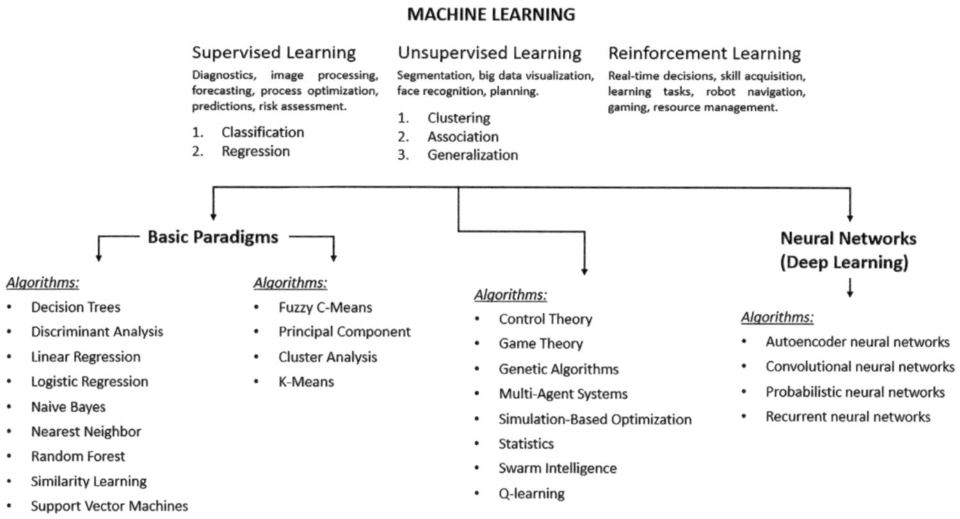

Figure 9.
Examples of machine learning algorithms [23].

weighted combination based on training data that can be used later to deduce results for the incoming data. **Figure 9** presents examples of the mostly spread machine learning algorithms in the diagnosis of electrical machines [23].

One of the primary drawbacks of machine learning algorithms is that they need a lot of data to train a high-accuracy model. However, it usually depends on the complexity of the model. Suppose the model will be used for classification, with classification being divided into two labels. In that case, the accuracy will be pretty high even with a low training data set. But suppose that is to be changed by classifying the classification into four different labels. In that case, the system's complexity will increase, resulting in the algorithm needing more data to make an accurate model. **Figure 10** shows the general working of a machine learning or neural network model, to be precise.

There are different types of machine learning algorithms based on specific logic. The training data set results in a statistical complex function based on the selected algorithm that gives a trained model. Among the machine learning algorithms, the most used are neural networks. Neural networks are further divided into three main types:

- artificial neural network (ANN),
- recurrent neural network (RNN),
- convolution neural network (CNN).

ANN and RNN are primarily used for training for models related to detection or prediction. Most ANN models are regression-based or feed-forward models, whereas RNN is feed backward neural network models. Neural network model training is divided into three layers: the input layer, the hidden layer, and the output layer. The hidden layer is where the weighted nodes are set up, as the weight of these nodes is adjusted with each training data set. Once the model is trained using the training data set, a blind validation can be carried out to test the accuracy of the model before implementing it in a real-time scenario.

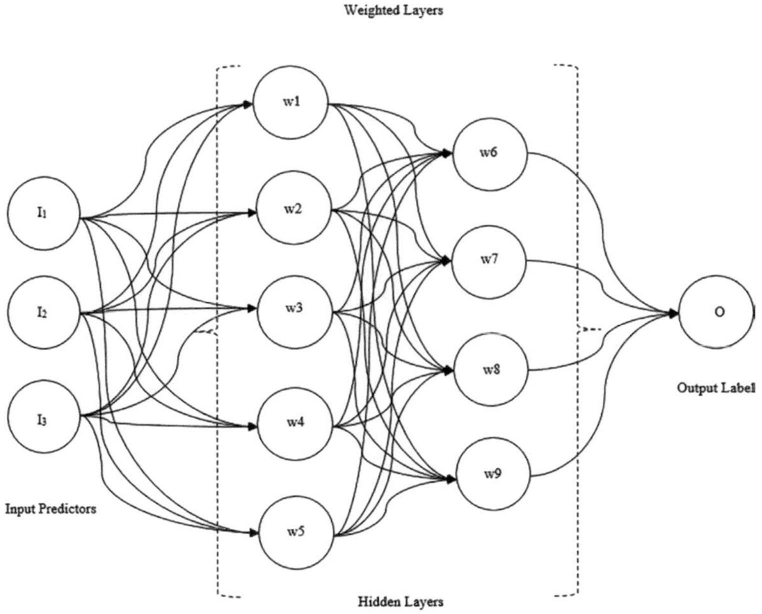

Figure 10.
Neural network schematics.

These models, after training, are usually implemented in the cloud and are used to detect faults in the incoming signals. Although they can be trained to be precise, the data needed for it is usually great. That is why researchers are looking into generating such data programmatically based on the real-time collected data and frequency harmonics. If this is reached, it will be possible to mass produce faulty data according to the need of the electrical machine. This will help train a machine learning model specifically for the required scenario. Implementing these models on nodes or edge devices might also be possible. This will help move from cloud computation toward edge computing.

Training of machine learning models also has other issues with accuracy, based on the complexity of the system. **Table 2** shows the comparison of accuracies for two different types of training sets having different labeled categories. The comparison

Neural network algorithm	Smaller training set		Bigger training set	
	Two categories (%)	Four categories (%)	Two categories (%)	Four categories (%)
Narrow Neural Network	88.30	65.00	70.00	38.50
Medium Neural Network	82.50	63.30	73.50	46.50
Wide Neural Network	88.30	73.30	76.20	51.50
Bilayered Neural Network	82.10	62.10	75.20	43.10
Trilayered Neural Network	95.40	64.20	73.3	53.40

Table 2.
Accuracy comparison of different neural network models.

confirms that changes in complexity or size of training sets do impact the accuracy of different machine learning models. In this specific example, the algorithms were run with specific conditions to compare them under similar training and validation processes. However, the results can still be optimized as the training process (i.e., epochs, etc) and the test approach (i.e., v-fold cross-validation, holdout validation, etc) can also result in different results. Hence, changing approaches can result in better or, even in some cases, worst performance, for example, a trilayered neural network with two categories and a smaller training set can result in an overfitted model.

As the system becomes more complex, a larger number of data is needed, but this also shows that there is a chance that another machine learning algorithm can perform better for the same scenario. Hence, these trained models are still flexible and there is a need to either get the optimal number of data sets for the training of the machine learning-based models or implement a custom-made machine learning model that can help identify faults related explicitly to electrical machines with high accuracy.

5. Predictive maintenance

As the industry is moving toward predictive maintenance from scheduled maintenance, there is still much research to be done in this area. Most of the research going on is related to fault detection rather than fault prediction, but companies are working in this specific area. The most important thing in this field is to identify the faulty frequency components in the early stage of the fault and the behavior of the signal and its frequency components when the signal is shifting from a healthy state to a faulty state. Once these things are identified, the next step is to train such a model that will be able to predict whether the fault is going to occur and in how much time. This will depend based on pre-processing of data and classification of the components. This is not a small task and needs dedication and time.

Researchers are looking for better ways to get a prediction model for faults help to identify them even before they occur. This leads us toward predictive maintenance, there might be some companies that are already running some kind of predictive maintenance algorithms with their systems, but at the moment, the hardware setup they have to use alongside it is quite expensive. So, another main issue in this area is to make it such that it is not only cheap but also no specific hardware setup is needed in this regard. There are also multiple directions in which predictive maintenance trained algorithms can be utilized. There can be a combination of different algorithms to get higher accuracy or more accurate results. Similarly, fuzzy logic systems can also be used in accordance with machine learning algorithms and signal processing to get a more accurate system for predictive maintenance.

Another issue that the researchers commonly face in this aspect is the lack of data. As the data collected in an industrial environment is limited, especially in the case of faults, training a model with quality data and testing it out is quite difficult. Also, the data required to train models properly should be good in quality and quantity. Some researchers are working on observing the pattern in different faults to generate a statistical equation for the faults so that synthetic signals can be generated, which can help cover up this issue. The main issue in this aspect is to correctly identify the range of amplitude of frequency components that are generated when a fault is present in the electrical machine. This is not easy as it requires much data analysis and robust testing, but immense research is taking place in this direction.

6. Conclusion

The industrial revolution and information technology advancements have opened up new research areas to make things more convenient for industrial applications. IoT, with its usage in condition monitoring, fault detection, and remote controlling, is already becoming the norm for the industry. It will be more important in the near future to implement predictive maintenance for the industry to move away from scheduled maintenance to cut short on losses. Hence, fault diagnostics and predictive maintenance are the need of the hour. Here, a concise overview of a condition monitoring system is given along with the issues in the machine learning algorithm and the possibilities of predictive maintenance are discussed.

Although there are still many limitations, such as microcontroller boards are still in development, fault prognostics, limitation of available data, and lack of statistical and predictive models. However, much research is being done in these areas, with the microcontroller boards being advanced rapidly, making them more reliable and stable. An increase in their computational power will also result in a more stable and quicker transmission of data. The bigger problem is still related to the lack of data, resulting in trained models not being up to the mark. However, researchers are currently developing statistical models by reengineering. By observing the signals for different faults from an electrical machine, researchers are trying to develop statistical models that can generate signals similar to the fault. Although the process takes much time and concentration, researchers are getting near and it might be possible in the near future to generate faulty signals based on statistical models.

This chapter discusses a specific version of a condition monitoring system with a discussion related to maintenance approaches, machine learning algorithms, and some of the issues faced in this aspect.

Acknowledgements

This research was funded by [EEA]/ [Norway] Grants 2014–2021, "Industrial Internet methods for electrical energy conversion systems monitoring and diagnostics".

Author details

Hadi Ashraf Raja*, Karolina Kudelina, Bilal Asad and Toomas Vaimann
Tallinn University of Technology, Tallinn, Estonia

*Address all correspondence to: haraja@taltech.ee

References

[1] Asad B, Vaimann T, Belahcen A, Kallaste A. Broken rotor bar fault diagnostic of inverter fed induction motor using FFT, Hilbert and Park's Vector approach. In: XXIIIrd International Conference on Electrical Machines (ICEM'2018). Alexandroupoli - Greece September 3-6; IEEE. 2018. pp. 2352-2358

[2] Vaimann T, Sobra J, Belahcen A, Rassõlkin A, Rolak A, Kallaste A. Induction machine fault detection using smartphone recorded audible noise. IET Science Measurement and Technology. 2018;**12**(4):554-560. DOI: 10.1049/iet-smt.2017.0104

[3] Vaimann T, Kallaste A, Kilk A. Sensorless detection of induction motor rotor faults using the Clarke Vector Approach. Science Journal of Riga Technical University. 2011;**28**(1):16-21

[4] Sathyan S, Belahcen A, Kataja J, Vaimann T, Sobra J. Computation of stator vibration of an induction motor using nodal magnetic forces. In: 2016 XXII International Conference on Electrical Machines (ICEM), 4-7 September. Lausanne, Switzerland: IEEE. 2016. pp. 2198-2203

[5] Kudelina K, Asad B, Vaimann T, Rassõlkin A, Kallaste A, Khang HV. Methods of condition monitoring and fault detection for electrical machines. Energies. 2021;**14**(22):7459. DOI: 10.3390/en14227459

[6] Barksdale H, Smith Q, Khan M. Condition monitoring of electrical machines with internet of things. In: IEEE SOUTHEASTCON. 19-22 April 2018, St. Petersburg, FL, USA. pp. 1-4. DOI: 10.1109/SECON.2018.8478989

[7] Parpla R, Lacob R. Application of IoT concept on predictive maintenance of industrial equipment. In: 8th International Conference on Manufacturing Science and Education – MSE 2017. Sibiu, Romania. 2017. pp. 1-8. DOI: 10.1051/matecconf/201712102008

[8] Mohanraj M, Thottungal R, Jaikumar K. A can bus based fault monitoring and diagnosis system in wind turbine. In: 2013 International Conference on Emerging Trends in VLSI, Embedded System, Nano Electronics and Telecommunication System (ICEVENT); Tiruvannamalai, India. 7-9 Jan 2013. pp. 1-3

[9] Wymore ML, Van Dam JE, Ceylan H, Quai D. A survey of health monitoring systems for wind turbines. Renewable Sustainable Energy Review;**52**:976-990

[10] Mukherjee S, Ghosh A, Sarkar SK. Arduino based wireless heart-rate monitoring system with automatic SOS message and/or Call facility using SIM900A GSM Module. In: 2019 International Conference on Vision Toward Emerging Trends in Communication and Networking (ViTECoN); Vellore, India. 30-31 Mar 2019. pp. 1-5

[11] Edwan E, Abu-Musameh M, Alsabah A. Blood pressure monitoring using arduino-android platform. In: 2020 International Conference on Assistive and Rehabilitation Technologies (iCareTech). Gaza, Palestine; 28-29 Aug 2020. pp. 87-91

[12] Khuwaja AA, Sattar A. Solar power remote monitoring and controlling using Arduino, LabVIEW and web browser. In: 2015 Power Generation System and Renewable Energy Technologies (PGSRET); Islamabad, Pakistan. 10-11 June 2015. pp. 1-4

[13] Janeera DA, Poovizhi H, Sheik Haseena SS, Nivetha S. Smart embedded

framework using Arduino and IoT for real-time noise and air pollution monitoring and alert system. In: 2021 International Conference on Artificial Intelligence and Smart Systems (ICAIS); Coimbatore, India. 25-27 Mar 2021. pp. 1416-1420

[14] Xie P et al. Sensing wind for environmental and energy applications. In: 25th IET Irish Signals & Systems Conference 2014 and 2014 China-Ireland International Conference on Information and Communications Technologies (ISSC 2014/CIIT 2014); Limerick, Ireland. 26-27 Jun 2014. pp. 64-69

[15] Mabrouki J, Azrour M, Dhiba D, Farhaoui Y, Hajjaji SE. IoT-based data logger for weather monitoring using arduino-based wireless sensor networks with remote graphical application and alerts. Big Data Mining and Analytics. **4**:25-32

[16] Sharma P, Kapoor SR. Evaluation of arduino based das for condition monitoring of induction motor. In: 2017 International Conference on Information, Communication, Instrumentation and Control (ICICIC); Indore, India. 17-19 August 2017. pp. 1-4. DOI: 10.1109/ICOMICON.2017.8279084

[17] Andrei O, Ungureanu S, Miron A, Cziker AC. IoT power monitoring device using Wi-Fi and Arduino. In: 2021 9th International Conference on Modern Power Systems (MPS); Cluj-Napoca, Romania. 16-17 Jun 2021 pp. 1-6

[18] Raja H, Vaimann T, Rassõlkin A, Kallaste A, Belahcen A. IoT based tools for data acquisition in electrical machines and robotics. In: 2021 IEEE 19th International Power Electronics and Motion Control Conference (PEMC); Gliwice, Poland. 25-29 Apr 2021. pp. 737-742

[19] Ahmad N, Bajwa SIS, Ahmad S, Kha L, Malik TA, Tahir Y. Autonomous unmanned surface vehicle a microcontroller based approach. Wuhan, China; In: 2011 International Conference on Electric Information and Control Engineering (ICEICE 2011). Wuhan, China. 15-17 Apr 2011

[20] Kuts V, Otto T, Tähemaa T, Bukhari K, Pataraia T. Adaptive industrial robots using machine vision. In: ASME International Mechanical Engineering Congress and Exposition, Proceedings (IMECE); Pittsburgh, Pennsylvania, US. 9-15 Nov 2018. DOI: 10.1115/IMECE2018-86720

[21] Kuts V et al. Synchronizing physical factory and its digital twin throughan iiot middleware: A case study. Proceedings of the East Academic Science. 2019; **68**(4):364-370. DOI: 10.3176/proc.2019.4.03

[22] Debbag Y, Yilmaz EN. Internet based monitoring and control of a wind turbine via PLC. In: 2015 3rd International Istanbul Smart Grid Congress and Fair (ICSG); Istanbul, Turkey. 29-30 Apr 2015

[23] Kudelina K, Vaimann T, Asad B, Rassõlkin A, Kallaste A, Demidova G. Trends and challenges in intelligent condition monitoring of electrical machines using machine learning. Applied Sciences. 2021;**11**(6):2761. DOI: 10.3390/app11062761

Chapter 2

Perspective Chapter: Data Mining in Electrical Machine Maintenance Reports

Karlis Athanasios, Falekas Georgios, Verginadis Dimosthenis and Jose A. Antonino-Daviu

Abstract

Industrial electrical machine maintenance logs pertinent information, such as fault causality and earlier indications, in the form of a semi-standardized report, previously written and now in digital form. New practices in predictive maintenance, state-of-the-art condition monitoring, include increasing applications of machine learning. Reports contain a large volume of natural text in various languages and semantics, proving costly for feature extraction. This chapter aims to present novel techniques in information extraction to enable literature access to this untapped information reserve. A high level of correlation between text features and fault causality is noted, encouraging research for extended application in the scope of electrical machine maintenance, especially in artificial intelligence indication detection training. Furthermore, these innovative models can be used for decision-making during the repair. Information from well-trained classifiers can be extrapolated to advance fault causality understanding.

Keywords: artificial intelligence, big data applications, computer aided engineering, condition monitoring, deep learning, electrical machines, industry 4.0, knowledge acquisition, predictive maintenance, supervised learning, text mining

1. Introduction

Maintenance of electrical machines (EMs) follows industry-established procedures according to subject type and encompasses measurements, maintenance efforts' results, and technician expertise, all of which are necessarily logged in what is called a maintenance report (MR). Widespread information storage forms include numerical and audiovisual data. However, this data is almost always accompanied by a natural text (NT) in the industry's regional language form, to provide context. Each fault typically follows a similar pattern, observations of which are logged in the sequential flow or keywords in the natural text, containing information about degradation procedures, fault causalities, and any similar relevant comments. These semantics are naturally followed and produced by human cognition, effortlessly granting the reader understanding about the fault type and solution, generating patterns.

Numerical data is the main contribution of an MR. Values of interest are tracked according to well-established sensor hardware configuration standards and expertise. However, a standalone number is not enough to assess equipment condition. For example, low stator winding insulation ohmic resistance could indicate either particle contamination or thermal aging [1]. Additional information is required to collapse this quantitative observation into a qualitative one. The industry tends to be the easiest additional experiment, which in this case is a visual inspection (VI). A logical avenue to log such information is in one of three ways: videos or pictures, evidential objects, and NT. Since insertion of observation in the already existing report form as NT is the easiest, fastest, and cheapest solution, it is almost always preferred. Evidence and media are preserved for more dire situations and only when necessary.

NT is a convenient medium for human cognition. We are naturally trained to extract semantics and information from a high redundancy medium, our language, spoken or written. Furthermore, parallels and conclusions regarding causality and precedence can be drawn seamlessly, and transferred via the critical assessment of the responsible engineer to the interested party. These correlations are often made without conscious effort and therefore vary throughout. Thus, expertise is realized and advanced throughout the years.

The advent of Industry 4.0 has brought novel tools and techniques capable of expanding the understanding, processing, and handling of systems and procedures. Heralding the emergence of Big Data, interconnectivity, and digitization, each observation or case study can aid in establishing accurate correlations and training science's newest and sharpest tool, Artificial Intelligence (AI). AI relies on copious amounts of information in the form of input–output numeric pairs. With this outlook, NT proves costly for statistical processing, resulting in a significant volume of unused data.

Natural Language Processing (NLP) is an interdisciplinary sub-field of computer science, artificial intelligence, and linguistics, with the aim to quantify semantics and information in NT. NLP techniques are increasingly being investigated in broader literature for processing and understanding this data. Applications are rapidly multiplying in the past 5 years, facilitated by combined efforts and new hardware. These now robust methodologies are ready to be investigated and applied in numerous fields, such as EM condition monitoring (CM). In combination with traditional numerical causality analysis, graphs depicting common patterns and decision trees can be composed [2]. Case studies are of paramount importance to not only aid in the optimization of concurrent techniques by means of additional input, but also draw attention and confirm results.

Specifically concerning EM CM, such an attempt has not been extensively researched in literature, to the best of the author's knowledge. Conceptually, this discipline provides two important facilitations: extensive expert knowledge and established procedures. This chapter aims to further the establishment of an innovative concept for NLP in the environment of industrial EM CM. A vision of this work is synergy between experts and machinery, in the form of understanding the context of observations, asking for further information, and then providing a verdict, which is the typical procedure undertaken in the industry with the limitations explained above. AI can automate report cognition and event causality graph production. The authors consider that this endeavor will aid the acceleration of PM in the new industrial paradigm by enabling access to a previously inaccessible vault of information. An overview of similar endeavors in a broader context can be found in [3].

2. Standard diagnostic procedure

EMs are the primary mobilization force of industry and electricity generation especially. **Figure 1** presents global electricity production [5]. Apart from solar and other renewables, the rest of the sources are utilized as kinetic sources for turning the rotor of a synchronous generator (SG), or an induction machine (IM) in the case of newer wind turbines. Therefore, 95% of global energy includes SGs in its production chain.

Hence, proper operation with minimal losses and downtime is of paramount importance. Even operation under sub-optimal power factor -increased losses- should be avoided, as a considerable amount of energy is wasted. EMs consist of a multitude of electromechanical parts which can be ailed by various faults, with their severity ranging from minimal (power factor reduction) to catastrophic (destructive failure). CM tackles the possibility of these faults by their statistical order of appearance. **Figure 2** depicts fault distributions in EMs [6]. Insulation faults represent the highest share in large industrial SGs and therefore attract research focus, followed by the bearings. Rotor faults present a universal, constant appearance. The cited research agrees with similar surveys done by EPRI and IEEE [7].

Therefore, the EM insulation system and especially that of the stator must face and withstand various faults while remaining reliable for the EM to stay healthy and optimal in its operation. The typical EM stator is similar in both SGs and IMs and consists of [8]:

- the copper conductors, which must have a large enough cross-section in order to carry the required current without overheating;

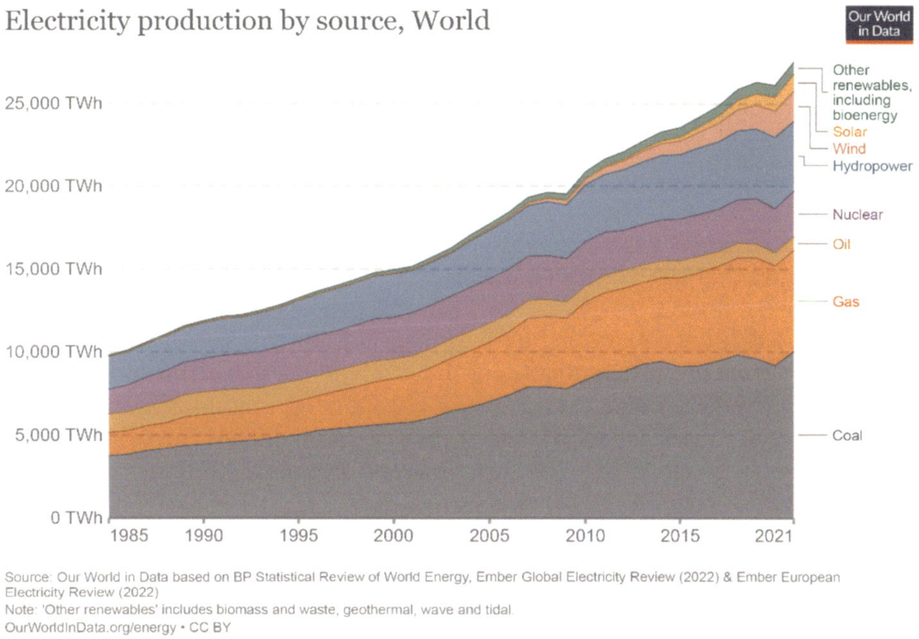

Figure 1.
Global electricity production by source. Based on the primary source, 95% of global production utilizes mostly synchronous generators [4].

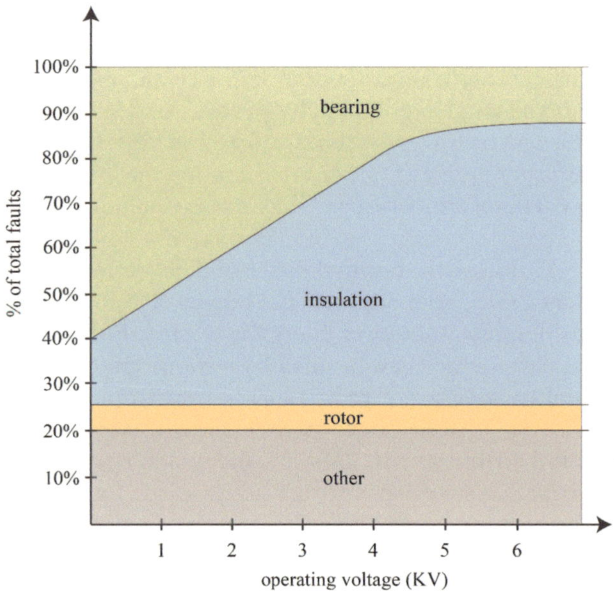

Figure 2.
Fault distribution in IMs according to their operating voltage level. As the voltage level increases, insulation faults become more prominent. The same holds for SGs.

- the stator core, which consists of thin sheets of magnetic steel and prevents the flowing current in adjacent conductive material;
- the electrical insulation, which is passive but necessary for the EMs and consists of:
    - Sub conductor insulation;
    - Turn/strands insulation;
    - Ground wall (main) insulation.

Constant or transient stresses can affect the insulation deterioration of an EM. These stresses are thermal, electrical, mechanical, and ambient, commonly known as TEAM stresses.

Thermal stress [9] is determined by the losses within the conductors and plays the most significant role in the degradation of the insulation system. The operating temperature of the windings, which is a result of I^2R, eddy current, load losses, and heating due to core losses, is the primary source of thermal stress. Increased temperatures lead to more frequent chemical reactions and insulation lifetime is described by using the Arrhenius law:

$$L = Ae^{B/T} \tag{1}$$

where L is lifetime, T is the absolute temperature, A is the frequency of molecular encounter and $B = -E/R$, where E is the activation energy, which is constant for a given reaction and R is the universal gas constant. Thermal stress results in high-temperature differentials, overload, and hot spots.

Electrical stress [10] is related to the thickness of the electrical insulation of EMs.. It can lead to partial discharges (PD), which are small electric sparks that occur within air pockets of the insulation or on the surface of coils. In this case, the lifetime of the insulation is described by:

$$L = cE^{-n} \tag{2}$$

where L is lifetime, c is a constant, E is the stress level (kV/mm) and n is the power law constant. Surges, overvoltages, and partial discharges are indicative consequences of electrical stress.

Ambient stress [8] is caused by miscellaneous factors that can lead to stress and typically amplify main stress categories via their mechanisms. Ambient stress sources include moisture on the windings, presence of oil or dust, high humidity, broken particles within the EMs, and aggressive chemicals. The results can be aggressive and reactive chemical reactions degrading the machine parts, as well as contamination.

Mechanical stress [9] is caused by force acting on the parts, stemming from mechanical vibrations or electromagnetic forces, such as end-winding vibrations. The lifetime of the insulation is described using:

$$L = D\sigma^{-m} \tag{3}$$

where L is the lifetime, D, m are constants related to the insulation material and σ is the applied stress in $N/_{mm^2}$. There is vibration and oscillation in slot sections and in end winding.

Deterioration of the insulation system is typically caused by two or more stress factors that are responsible for that specific result. Multiple stresses both accelerate and evolve the failure, leading to more significant problems.

Various diagnostic tests can be used for the evaluation of the condition of an EM. These tests are undertaken after the EM has been manufactured, installed on-site, during periodic maintenance checks, or when fault indications occur. Standard offline diagnostics follow a common sequence, which utilizes the following categories of experiments.

2.1 Visual inspection (VI)

VI is the standard and usually the first offline diagnostic procedure, because it gives information for most possible faults both on the stator and rotor, indicating the necessity of further testing. VI utilizes a borescope, which is an optical device consisting of a rigid or flexible tube with an eyepiece or display on one end and an objective camera on the other, linked together by an optical or electrical system in between. A typical borescope is shown in **Figure 3** [11]:

- A flexible or a rigid tube
- An eyepiece
- A light source
- Optical lenses

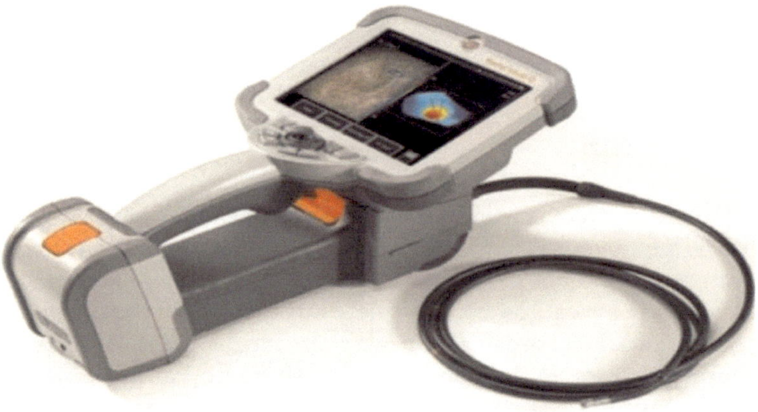

Figure 3.
Commercial typical Borescope [11].

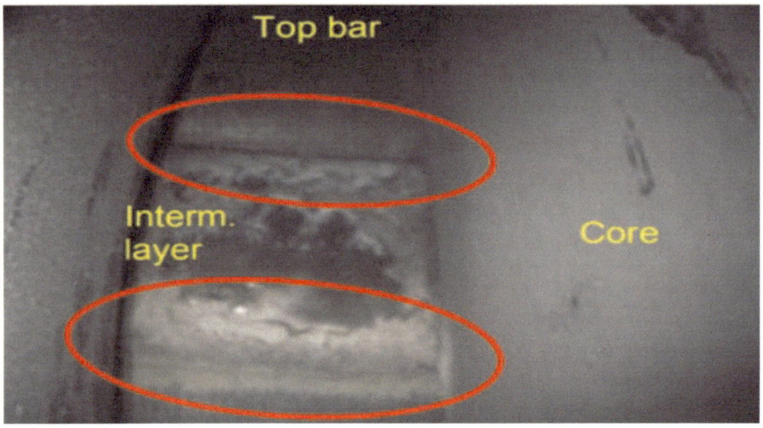

Figure 4.
Mechanical erosion.

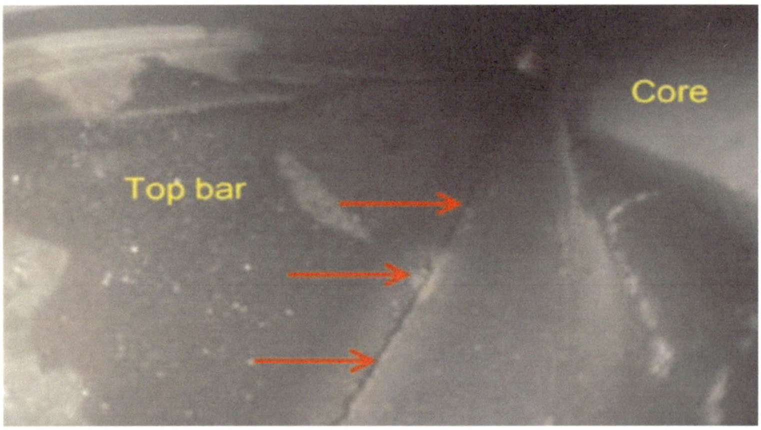

Figure 5.
Electrical erosion.

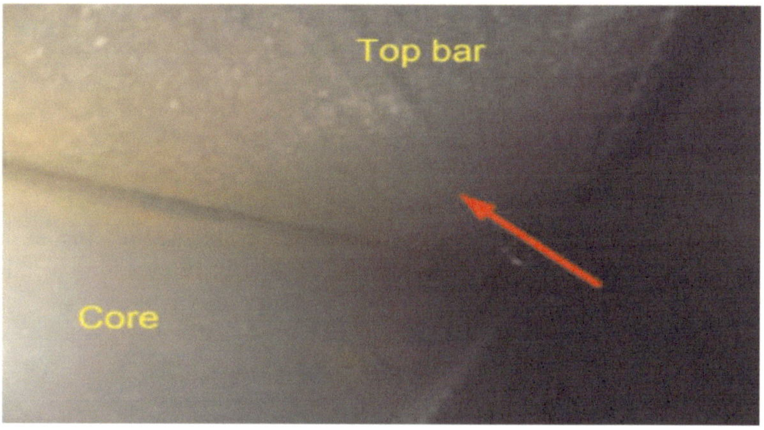

Figure 6.
PD effects.

Figure 7.
Broken Bar.

Common goals are the detection of partial discharge spread on bars and resulting mechanical erosion. It is performed through core ventilation channels from inside the cooler channels. VI is also suitable for detecting humidity, thermal and mechanical deterioration, cracks, ground wall insulation, insulation degradation, and turn-to-turn failures.

The following figures show several problems detected by using a borescope inside the stator of a real SG. Specifically, **Figure 4** shows significant mechanical erosion at the top bar of the SG, **Figure 5** shows the effects of electrical stresses, **Figure 6** shows electrical erosion at the bar due to PDs, and **Figure 7** shows cracks at the bar of the SG. These images may suggest following electrical testing.

2.2 Insulation resistance (IR) and polarization index (PI) test (Std. IEEE 43: 2013)

Indicative quantities are depicted as the IR value denoted by the time of calculation (in minutes) after the application of a voltage source on the ends of the insulation component. Typical values include R_1 = 100 Ω for most AC windings built after 1970 and R_1 = 5 for most EMs with random-wound stator coils and form-wound coils rated below 1 kV and DC armatures [9, 12].

PI tests follow the same standard with IR. Minimum PI = 1.5 for thermal class A (105) and minimum PI = 2 for class B (130) and above.

The two previous tests are always performed together -due to their dependence, explained below- in what is called the IR/PI test, which is suitable for checking machine windings for insulation deterioration. These tests are the most common for detecting potential problems in windings caused by contamination and pollution. Moreover, humidity, moisture, and cracks can be detected with these tests. PI can also determine if there is thermal deterioration.

The IR/PI test is done right at the machine terminals, one phase at a time, with cables and transformers disconnected. A high-voltage DC supply and a sensitive ammeter are required. The IR test measures the resistance of the electrical insulation between the copper conductor and the core of the stator or rotor. The PI is defined as the ratio between the IR measured after the voltage has been applied for 10 minutes and IR measured after 1 minute. Both IR and PI values decrease as an EM operates over the years because of the inevitably higher pollution penetration into the EM's windings. This means that IR is both initially reduced as time progresses and is also less resilient to the constant thermal stress applied by the current. IR is based on Ohm's law:

$$R = \frac{V}{I} \tag{4}$$

where V is the applied voltage and I is the sum of capacitive current, conduction current, leakage surface current, and absorption current. Lower IR is an indication that a problem exists within the insulation system since resistance has been lowered by contaminants or defects.

PI is a variation of the IR test. PI is the ratio of the IR measured after the voltage has been applied for 10 minutes (R_{10}) to the IR measured after just 1 minute (R_1):

$$PI = \frac{R_{10}}{R_1} \tag{5}$$

2.3 Measurement of power factor (PF) and dissipation factor (DF) (Std. IEEE 286: 2000)

According to the aforementioned standard and common knowledge, PF can be between 0 and 1. The same goes for DF, according to IEC 60034–27.3. Dissipation and power factors provide an indication of the dielectric losses within an insulation system. These measurements are conducted to identify if there are variations in C, DF, and PF over time, which indicate partial discharges or insulation degradation.

The DF is measured with a balanced bridge-type instrument, where a resistive-capacitive network is varied to give the same voltage and phase angle (tan delta) as measured across the stator winding. The DF is calculated from the R and C elements in the bridge that give the null voltage. This test is used for detecting humidity, moisture, PD, dielectric losses, and insulation degradation [13] (**Figure 8**).

The PF is measured by accurately measuring the Voltage applied between the copper and the core of a winding and detecting the resulting current. Also, it is necessary to measure the power of the winding with a wattmeter. So, the PF is:

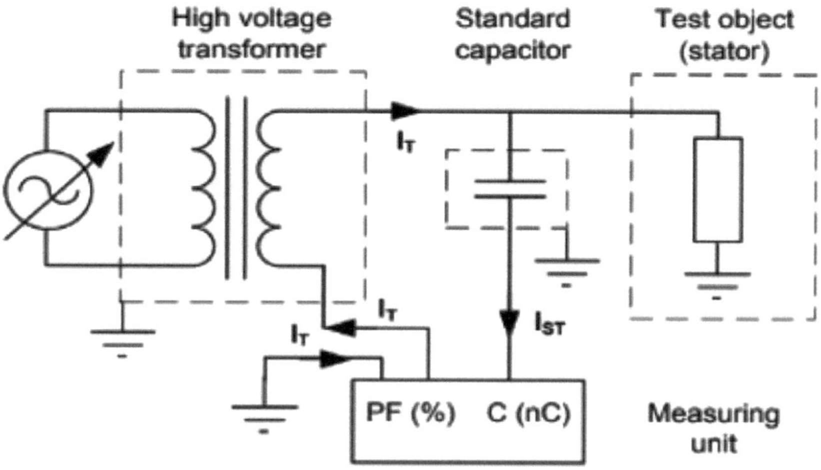

Figure 8.
PF/DF measurement scheme.

$$PF = \frac{W}{VI} \tag{6}$$

Comparing the two methods, the PF test is less accurate but less expensive as there is no need for a bridge-type instrument. The measurement of the DF can give information about PD activity, and contamination, while PF cannot. The DF can be converted to PF using:

$$PF = \frac{DF}{\left(1 + DF^2\right)^{0.5}} \tag{7}$$

The measurements of PF can be used for detecting possible problems and faults, such as humidity, moisture, overheating, dielectric losses, and insulation degradation. A wattmeter is used in order to measure the power to the winding and a voltmeter for measuring the applied voltage between the copper and the core of the winding and detecting the current.

2.4 Impedance test (Std. IEEE 112: 2004)

Humidity, moisture, thermal and mechanical deterioration, insulation degradation, and turn-to-turn failures are the faults, which can be detected by the measurements of the impedance. An AC source is used and different values of current are used in order to investigate the corresponding voltage values. Then, the impedance is calculated by using Ohm's law [14]:

$$Z = \frac{\Delta V}{I} \tag{8}$$

Figure 9 shows the results of an Impedance Test on a real Syncronous Generator. The four different lines are explained above the diagram. SR1 indicates the impedance of slip ring 1, while SR2 indicates the impedance of slip ring 2.

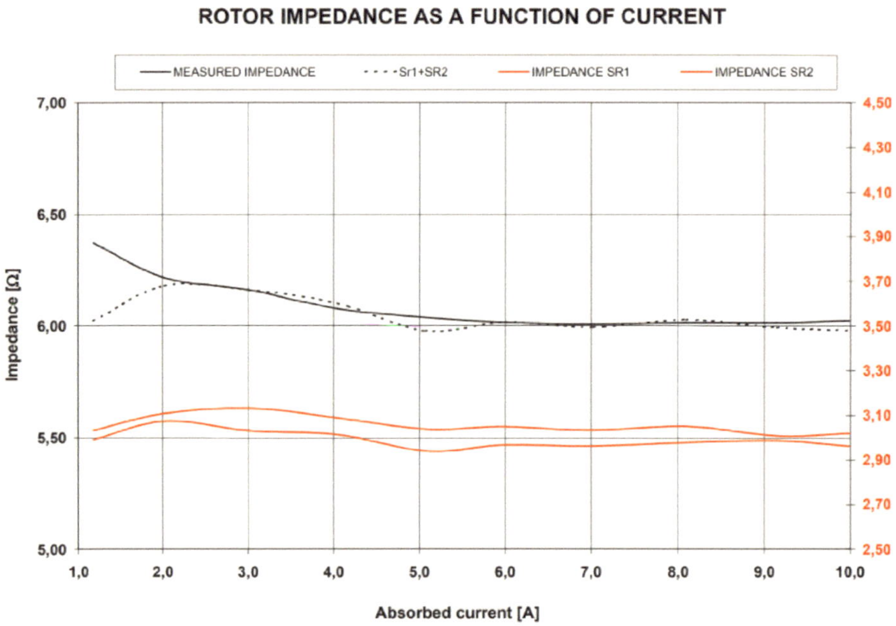

Figure 9.
Impedance test results.

2.5 Recurrent surge oscilloscope test (RSO) (Std. IEEE 56: 2016)

A low-voltage and high-frequency surge wave is injected at each slip ring. The test is based on an oscillograph inspection of the voltage traveling wave between the slip rings along the symmetrically constructed winding field rotor. A low voltage high frequency surge wave is injected at each one of the slip rings. The two signals are then compared to determine if the same waveform is observed at each slip ring. If the waves are identical then no short circuits are present. Variations in the pattern of the two waveforms would indicate shorts to be present. If the two signals have differences, interturn, ground, and turn-to-turn faults as well as insulation degradation are the possible faults. A power source, a reflectometer, or/and oscilloscope are appropriate for this test [15, 16].

A typical RSO diagram is shown in **Figure 10**. The two waveforms are identical and this means that the rotor is free of the aforementioned possible faults. It must be noted that SR1/OS indicates the voltage of slip ring 1 and SR2/IS indicates the voltage of slip ring 2.

2.6 Structure of a report

Every industrial EM has to be checked periodically by specialized technicians. After that, technical reports, which contain useful information and data for the inspection as well as the history of the inspected EM, are created and given to the industry in order to be informed about the condition of its EM and decide what actions must be taken. Moreover, reports like that can be used and analyzed by researchers in order to create prediction models for the condition of EMs. Specialized experience and real data from measurements on real EMs are elements that are

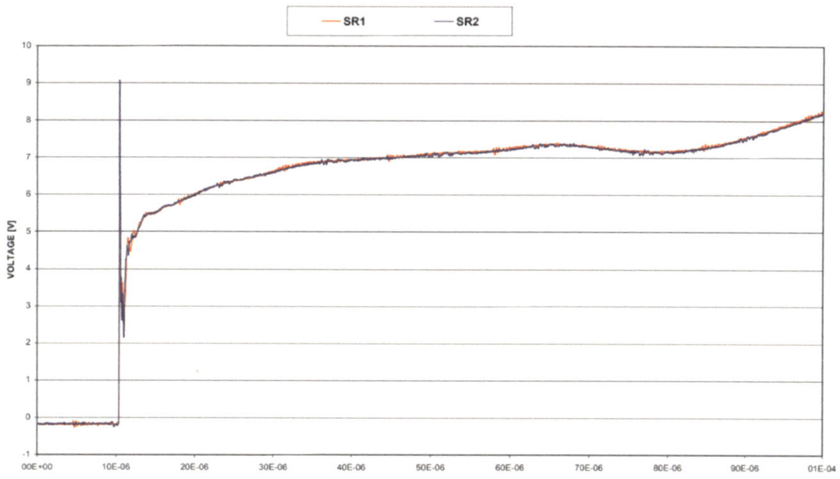

Figure 10.
RSO test results.

missing and they are very useful for creating prediction models with direct connection with the real situations of an industrial EM. Therefore, such reports are very significant for both industrial and research issues.

A typical structure of the reports used for the training of the proposed model of this manuscript is:

- Introduction - EM's Historical Issues - Milestones: The purposes of the Diagnostic Tests as well as information about the EM and significant dates are presented;
- Operation and Technical Data of the EM: rated power, voltage, current, frequency, power factor, dimensions, cooling type, number of poles, and other pertinent measurements;
- Selected Tests and Inspections: Different diagnostic methods are chosen each time according to the condition of the inspected EM, general considerations employed in selection;
- Results of the aforementioned tests and inspection: detailed information, data, diagrams, and pictures about the results of each diagnostic method as well as comparison with the previous years' diagnostics;
- Proposed maintenance actions according to the results of the measurements and information about the next date for diagnosis according to the results of the measurements.

Figures 11 and **12** highlight the different parts of a commonly found industrial SG report, created as a general template for the MRs studied during this research. The reports are considered as semi-structured, a term used in NLP to describe documents with structured information (tables, figures, lists etc.) interlaced with natural text. The structure within the report can be used to guide the AI processing it. For example, we know that the technical data table lists any number of parameters in any order,

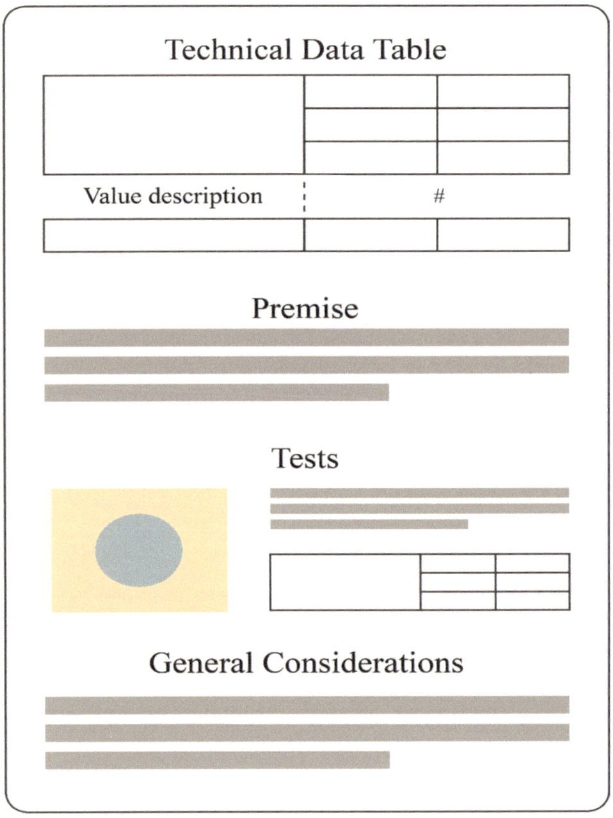

Figure 11.
First, more generic part of a typical SG maintenance report, just after the introduction.

therefore, the AI should expect a variable name partnered with a value. Furthermore, report language terminology is "specialized," meaning the total vocabulary is limited and populated with sector-specific terminology, further assisting by limiting the range of the employed interpretation.

As depicted in the pertinent figures, general information and considerations are commonly stored in just NT. The test section employs the most intuitive form of storage, such as imagery for VI. Actions and results are localized, meaning they are divided in different subsections for each component i.e., stator, rotor, and outside the frame. Actions are almost always depicted in lists (intuitively), while results follow the tests' paradigm of logical choice of medium. This localization can be employed in problem formulation to facilitate a correlation between cause and effect, while the list can provide the order of operations if logged properly.

3. Data mining reports using natural language processing

Processing information with logical processors or computer algorithms requires that its data is in numerical form. Furthermore, this form should also be in the appropriate context and facilitate necessary mathematical transformations. While this procedure is intuitive for numerical data such as measurements due to their

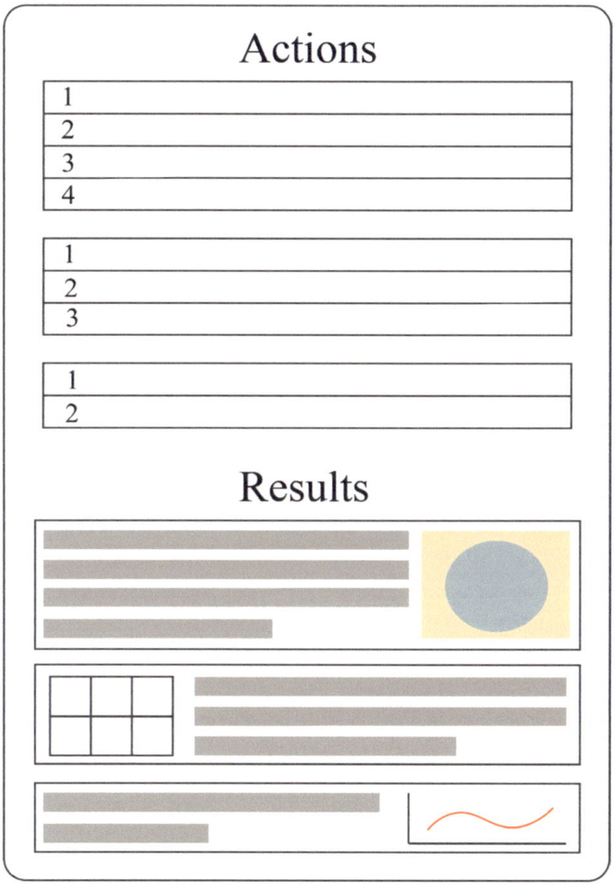

Figure 12.
Second, more specific part of a typical SG maintenance report.

underlying physical meaning and explored mechanisms, text processing requires a sophisticated approach.

The scientific field of NLP concerns itself with constantly improving existent and exploring new ways of information extraction from natural text. In the context of EM CM, it is important to understand the ways state-of-the-art procedures interpret text bodies, as the interpretation is closely tied to fault causality and can improve not only correlation understanding between different faults -since faults always are cascading- but also aid training of Industry 4.0's prodigal child, AI pattern recognition. This chapter introduces general principles of NLP state-of-the-art to aid the investigation of this promising avenue by discussing them in the context of EM CM. The successful application will enable tapping into a previously underutilized information source while paving the way for further storage, necessary under new industrial paradigms.

3.1 General principles

Language interpretation by machines is being explored for several decades, with the most successful principle rising to be word embedding. Understanding the embedding requires learning some preliminary machine learning (ML) concepts. The analysis presented hereafter includes CM examples.

Features are the input variables, here being sentences or words from the text body. Labels are the things to predict. For example, the sentence "the machine was found to have an increased PI value" could have the label "thermal degradation". A straightforward structure is having the words as features. Each selection as an input to the interpreter (could be a sentence or several, or a single word) is referred to as an Example. Examples can be labeled or unlabeled. It is important to understand that, as of now, all examples present in any report corpus are unlabeled. ML algorithms require labeled sentences to be trained. Therefore, an important issue to be solved is the production of labeled examples.

The training result, that is the prediction of a label given the features, is called the model. The model consists of the structure and weights of the classifier. A model can solve either a classification or a regression problem. Regression is continuous value prediction, while classification refers to discrete predictions. The most straightforward way to structure a fault prediction model would apparently be classification, such as the example provided above. However, regressive models could also prove to be useful and should always be considered.

An important metric in evaluating a training result -the weights and bias of the model- is Loss. Loss is a number indicating how bad the prediction of a single example performed. A perfect model would have zero loss, increasing with each failed prediction, and how far from the correct answer it is. Minimization of loss is therefore the function that the training is based upon and should be carefully formulated.

Low loss is not a complete indication that our model performed adequately. A model could have low loss but perform poorly when introduced with new unseen examples. This concept is described as overfitting, which is when our model has a poor generalization capability. Overfitting occurs when the model is more complex than necessary, proven time and time again and manifested in Ockham's razor. But how can we create a model from scratch based on a text corpus and provide it with unseen data for validation? The answer is separating the dataset into the training, validation, and test subsets. Proper separation is of paramount importance to the training, with existing ML paradigms found in the literature. Training on the same dataset, however, still exhausts and overfits the model. It generally is a great idea to keep refreshing the dataset with new reports while continually adjusting the established model.

As previously discussed, in order to enable the model to multiply features with their weights, said features should be numerical values. This process is called Feature Engineering and is a critical step in ML. In the case of text, the most straightforward way to map words is the so-called One-Hot Encoding. This encoding utilizes a vector of dimensionality equal to the vocabulary (total number of different words), where each word corresponds to a specific dimension of the vector, in which place its value is 1, with the rest being 0. Dimensionality can be reduced by aggregating infrequent words into an out-of-vocabulary (OOV) group. While one-hot encoding is the basic understanding example of word encoding, advanced techniques are preferred.

Now on to the gist of NLP. The next step up from proper encoding comes in the form of an Embedding. Shortly, an embedding is a vector in low-dimensional space coded such that the feature it represents is nearby similar ones. Take for example the words "king" and "queen." These words should be relatively close in the embedding space since they refer to a similar quality. Furthermore, an intuitive embedding would also support a mathematical operation, for example, "+", so that the calculation would be: "king" – "male" + "female" = "queen". A possible problem with utilizing the one-hot encoding in this example is that the dimensionality of a sentence would be arbitrarily large and sentence vectors would be very sparse; a lot of zeroes with very few ones in between. Therefore, proper embeddings are required. Thankfully, NLP research has provided tools for the task.

3.2 BERT: An interesting approach

Concerned with EM CM and NT information extraction specifically, a logical approach would be to use developed and established tools to achieve our goal, rather than making one from scratch. The choice of tools is largely based on their performance in established standards and engineering intuition -whether it suits the pertinent problem-. To that end, Bidirectional Encoder Representations from Transformers (BERT) are the tool of choice to investigate possible correlations in our task [17].

NLP state-of-the-art employs pre-trained language models. That is, these models employ a trained embedding model, with additional semantic ML analysis, quickly defined in the previous subchapter. It is important to remember that with ML tasks, time efficiency is of paramount importance. Due to the complexity and nuance of text mining, model training requires immense hardware capability and a time sink. Thus, pre-trained models have been extensively researched to be employed in final tasks.

Two general separations occur in possible applications: level and approach. The two levels are sentence and token levels. Sentence level includes the entirety of two or more sentences as input and attempts to predict their relationship with a holistic analysis. Token-level tasks provide a more precise output at the word level and are suited for question answering and named entity recognition. CM tasks are approached as sentence-level; we attempt to predict faults via sentence relationships.

The approach could be either feature-based or fine-tuning. A feature-based approach would be largely dependent on the task at hand; additional model architecture is designed specifically for the problem. Fine-tuning is a novel technique that utilizes the pre-training and keeps the same architecture with parameter training on the task at hand. Both approaches would be suited to CM and are up for debate. BERT utilizes fine-tuning.

One novelty provided by the BERT approach is its bi-directional representation. Previous state-of-the-art models would approach a sentence unidirectionally or at best aggregate the left-to-right and right-to-left representations. Consider the human interpretation of a sentence; we both speak and process information serially, or left-to-right but intuitively also consider the entirety of the sentence both in formulation and processing. Therefore, a bidirectional representation would theoretically paint a more complete image.

One more parameter to consider is whether the training is done supervised or unsupervised. BERT employs an unsupervised training approach due to the nature of semantics extraction from NT, that is to infer new possible correlations from the information contained within rather than the already established knowledge, which would render the research point moot. Pre-training approach also enables transfer learning, which transfers knowledge from larger datasets and/or supervised tasks. The natural language contains base semantics that apply to multiple different problems in different iterations. Consider the following example: a learned differentiation between "positive" or "negative" would also apply contextually to "healthy" or "faulty". This knowledge is contained within a broader dataset and can be fine-tuned on the task at hand.

As a tool, BERT focuses on the fine-tuning approach, making it essentially plug-and-play for our operation. Architecture appears to fit the CM context, while its performance based on important metrics is state-of-the-art. Therefore, it is promising for an initial attempt.

3.3 A deeper dive

Transformer [18] utilization presents a unified architecture across different tasks; both at the pre-training and fine-tuning steps, as well as the capability to be employed on multiple input forms such as images and video, allowing for a possible expansion of MR information extraction apart from its text corpus. The architecture allows for training on unlabeled data, which is paramount for the task at hand. While supervised, labeled fine-tuning is required and results in significantly improved performance, processing the entirety of the report corpus would prove immensely difficult. Deep consideration of experts is necessary for this endeavor in labeling examples to initialize training; processing unlabeled data is equally important in time efficiency.

The relationship between two sentences is not directly captured by language modeling, which is where the second stage of our training comes in, further realized at the task level by the fine-tuning mechanism. BERT's attention basis additionally allows for correlation between distant sentences, helping the endeavor. However, this transformer-based architecture may not be easily able to represent the entirety of the CM problem and purpose; while additional recurrent neural network (RNN) with long-short term memory (LSTM) neurons -extensively used in NLP- reduces typical BERT performance, additions may be required to fully encase the problem and have to be researched. Initial case studies with only BERT are being performed, pending judgment by experts. Overall, the underlying novel mechanism's benefits over older approaches can be summarized as:

- Faster and less complex representation;
- More interpretable models;
- Diversity in tasks;
- Behavior related to semantic and syntactic structure of sentences;
- Application to audio, video, and images.

3.3.1 Tokenization

Another interesting BERT component is the employed WordPiece embedding [19]. Due to its namesake, this embedding utilizes a limited vocabulary of sub-word units, further reducing dimensionality i.e., aggregating different forms of the same word. This procedure naturally handles the processing of rare words and is especially useful in semi-structured language corpora, such as our specialized engineering language.

The embedding includes a balance between character and word delimiters, enabling handling of newly seen (OOV) words with a completely data-driven approach and guaranteeing the generation of a deterministic segmentation for any possible sequence of characters. Additionally, an included length-normalization procedure and coverage penalty encourages complete coverage of source sentences in output. This essentially infinite vocabulary allows for open-ended optimization. However, this approach does not allow for observation of training errors with the built-in fitness function and requires a task-specific procedure to evaluate reward. Therefore, the proper definition of this function is important.

3.4 General approach

Having selected the tools at our disposal, briefly presented above, NLP research paired with ML paradigms and put in the context of the specific problem produces a general algorithm for this endeavor. However, before delving further in the discussion, a couple of disclaimers should be made. Firstly, the above tools have been evaluated and show promise, but researchers should remember that no approach is perfect. Interested parties are encouraged to test a diverse selection of NLP procedures in tackling the problem. Secondly, while generalized approaches do not provide specific solutions, the purpose of this chapter is the introduction of the proposed idea. Furthermore, the problem is open-ended and needs further elaboration in breaking down its complexity while setting standards. Thus, the proposed general representation is deemed appropriate. Finally, the explanation of BERT is kept simple in this chapter. Interested readers may refer to the open-source code and cited work for an in-depth, complete analysis. **Figure 13** presents an overview.

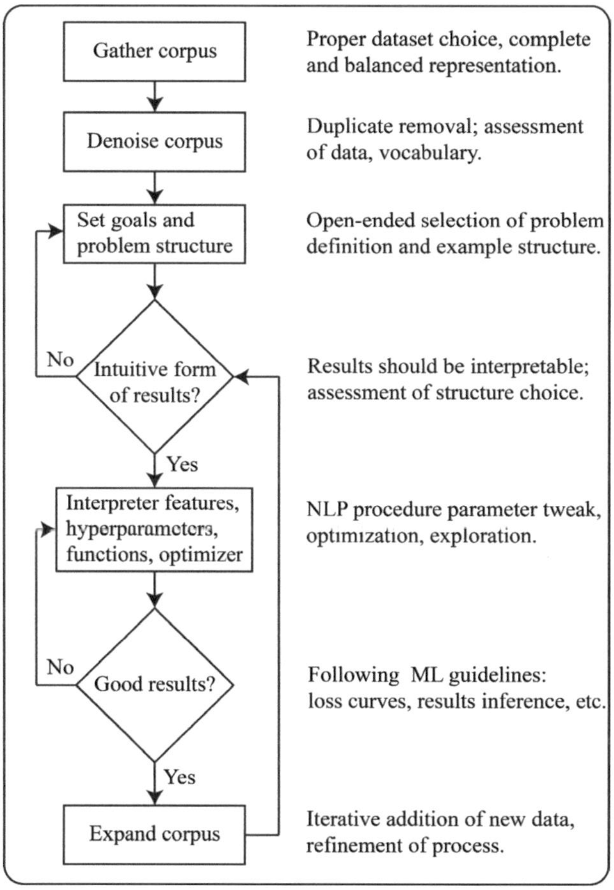

Figure 13.
General iterative process of machine learning problem establishment.

3.4.1 Discussion

While the proper selection of AI network architecture and hyperparameters is an important and difficult task on its own, the real challenge in this endeavor is structuring the problem. Were it only raw text processing, one would only look to combining NLP paradigms with EM CM expertise to formulate feature selection, perform labeling and optimize the procedure. However, one grand challenge in this endeavor is the robust processing of the reports due to their specific-but-varying structure, which is a double-edged sword; on one hand, this structure can be employed to better infer correlations and aid the interpreter with limitations; on the other hand, it should carefully be considered since the improper setting of the problem structure renders NLP impossible.

ML consensus agrees on an iterative approach. At first, a human manageable corpus is to be selected and denoised. After being deemed proper and balanced in its representation, the problem is set up with feature selection, labeling, and architecture choice, followed by optimization. When results are satisfying and intuitive, outside expertise should assess and offer an outside perspective; then, the process is expanded with new data, a classic procedure in ML tasks.

A human overview is the key to this research. In translation tasks, it has been reported to improve performance by up to 60% (according to pertinent test scores) [19]. It is important to understand that a complete AI CM is far off; this research aims to provide a tool for engineers to automatically and efficiently extract information from untapped datasets. EM MRs represent the intuitive knowledge of experts, which it attempts to quantify.

3.4.2 Dataset standards and default parameters

Finally, two initial steps in ML approaches are reaching milestones pertinent to dataset size, and finding initial parameters. This paragraph attempts to provide a few important values.

- BERT Model: largest allowed by hardware and time constraints;
- Mini-batch size: 16–48;
- Epochs: 2–3;
- Learning Rate: 2e-5 to 5e-5
- Sentences: >4.5 M;
- Vocabulary: 8-40 K;
- Optimizer: default ADAM [20].

Due to being open-source, there are numerous guidelines and examples of BERT applications. The above values have proven sufficient in the considered EM CM context via testing with real MRs, and should serve as a good starting point.

4. Conclusion

Data-driven approaches are substantially beneficial for new industrial and research paradigms such as Industry 4.0 and the emergence of Big Data. New methodologies, such as the Digital Twin [21], can greatly benefit from a large and structured database, especially in the context of EM CM, since faults are deeply correlated and their mechanisms are still partially obscured. This work presents a novel approach for structured data extraction from an untapped source of information, namely the knowledge stored in EM MRs in the form of NT. NLP is increasingly gaining traction due to the aforementioned circumstances and has not yet been employed in this field, to the best of the author's knowledge. We surmise that a breakthrough in this endeavor can greatly benefit the industry and attempt to initialize it with research in this department. This chapter serves as the overview of this attempt, providing extended knowledge acquired during related research.

Author details

Karlis Athanasios[1]*, Falekas Georgios[1], Verginadis Dimosthenis[1] and Jose A. Antonino-Daviu[2]

1 Department of Electrical and Computer Engineering, Democritus University of Thrace, Xanthi, Greece

2 Department of Electrical Engineering, Universitat Politècnica de València, Valencia, Spain

*Address all correspondence to: akarlis@ee.duth.gr

References

[1] Verginadis D, Antonino-Daviu J, Karlis A, Danikas MG. Diagnosis of stator faults in synchronous generators: Short review and practical case. In: Proceedings – 2020 International Conference on Electrical Machines. New York City, United States: ICEM; 2020. pp. 1328-1334. DOI: 10.1109/ICEM49940.2020.9270936

[2] Nakata T. Text-mining on incident reports to find knowledge on industrial safety. Annual Reliability and Maintainability Symposium (RAMS). 2017. pp. 1-5. DOI: 10.1109/RAM.2017.7889795

[3] Brundage MP, Sexton T, Hodkiewicz M, Dima A, Lukens S. Technical language processing: Unlocking maintenance knowledge. Manufacturing Letters. 2021;**27**:42-46

[4] Ritchie H, Roser M, Rosado P. Energy. Oxford, England. 2020. Published online: OurWorldInData.org. Retrieved from: https://ourworldindata.org/energy [Online Resource]

[5] Afrandideh S, Milasi ME, Haghjoo F, Cruz SMA. Turn to turn fault detection, discrimination, and faulty region identification in the stator and rotor windings of synchronous machines based on the rotational magnetic field distortion. IEEE Transactions on Energy Conversion. 2019;**35**(1):292-301

[6] Kammermann J, Bolvashenkov I, Schwimmbeck S, Herzog HG. Reliability of induction machines: Statistics, tendencies, and perspectives. In: The 31st IEEE International Symposium on Industrial Electronics (ISIE). New York City, United States: ISIE; 2017. pp. 1843-1847

[7] Swana EF, Doorsamy W. Investigation of combined electrical modalities for fault diagnosis on a wound-rotor induction generator. IEEE Access. 2019;7:32333-32342

[8] Stone GC, Boulter EA, Culbert I, Dhirani H. Electrical Insulation for Rotating Machines: Design, Evaluation, Aging, Testing, and Repair. Vol. 21. Hoboken, New Jersey: John Wiley & Sons; 2004

[9] Cimino A, Jenau F, Staubach C. Causes of cyclic mechanical aging and its detection in stator winding insulation systems. IEEE Electrical Insulation Magazine. 2019;**35**(3):32-40

[10] Saxén C, Gamstedt EK, Afshar R, Paulsson G, Sahlén F. A micro-computed tomography investigation of the breakdown paths in mica/epoxy machine insulation. IEEE Transactions on Dielectrics and Electrical Insulation. 2018;**25**(4):1553-1559

[11] Svensen M, Hardwick D, Powrie H. Deep neural networks analysis of borescope images. In: Proceedings of the European Conference of the PHM Society. Utrecht, The Netherlands; 2018. pp. 3-6

[12] Zoeller C, Vogelsberger MA, Fasching R, Grubelnik W, Wolbank TM. Evaluation and current-response-based identification of insulation degradation for high utilized electrical machines in railway application. IEEE Transactions on Industry Applications. 2017;**53**(3):2679-2689

[13] Bin Lee S, Younsi K, Kliman GB. An online technique for monitoring the insulation condition of AC machine stator windings. IEEE Transactions on Energy Conversion. 2005;**20**(4):737-745

[14] Verginadis D, Antonino-Daviu JA, Karlis A, Danikas MG. Determination of

the insulation condition in synchronous generators: Industrial methods and a case study. IEEE Industry Applications Magazine. 2022;**28**(2):67-77

[15] Kerszenbaum I, Maughan C. Utilization of repetitive surge oscillograph (RSO) in the detection of rotor shorted-turns in large turbine-driven generators. In: 2011 Electrical Insulation Conference (EIC). Vol. 2011. New York City, United States: EIC; pp. 398-401

[16] Imoru O, Mokate L, Jimoh AA, Hamam Y. Diagnosis of rotor inter-turn fault of electrical machine at speed using stray flux test method. In: AFRICON 2015. New York City, United States. 2015

[17] Devlin J, Chang MW, Lee K, Toutanova K. "BERT: Pre-training of deep bidirectional transformers for language understanding," NAACL HLT 2019 - 2019 Conference of the North American Chapter of the Association for Computational Linguistics: Human Language Technologies: Proceedings of the 2019 Conference. Seattle, Washington, USA. 2014. pp. 4171-4186

[18] Vaswani A et al. Attention is all you need. Advances in Neural Information Processing Systems. 2017;**2017**(Nips): 5999-6009

[19] Y. Wu et al., "Google's neural machine translation system: Bridging the gap between human and machine translation," 2016. pp. 1-23. ArXiv: abs/1609.08144

[20] Kingma DP, Ba JL. Adam: A method for stochastic optimization. In: 3rd International Conference on Learning Representations. San Diego, California, USA: ICLR; 2015. pp. 1-15

[21] Falekas G, Karlis A. Digital twin in electrical machine control and predictive maintenance: state-of-the-art and future prospects. Energies. 2021;**14**(18)

Chapter 3

Analysis of the Effects Produced by Pure Sine and Modified Sine Inverters in an Induction Motor

Arturo Yosimar Jaen-Cuellar, David Alejandro Elvira-Ortiz, Emmanuel Resendiz-Ochoa and Juan Jose Saucedo-Dorantes

Abstract

Most of the industrial applications are supported by complex machinery, which in turn are supported by electrical motors to perform specific tasks in multiple processes. Certainly, induction motors are the most widely used electrical machines in a majority of industrial machineries; in this sense, their operating condition plays an important role to ensure the machinery availability and to avoid unwanted stoppages. On the other hand, several sources may lead to producing faults in induction motors, but most of the common faults are produced by electrical or mechanical stresses, where the mechanical stresses are usually produced by unbalances or misalignments and the electrical stresses are generated by fluctuations or variations in the power supply. Thereby, when the induction motors are fed through inverters due to renewable energy, their operation may present slight variations since the sine wave has no perfect generation. In this regard, this work presents an analysis of the effects produced by pure sine and modified sine inverters in an induction motor. Such analysis consists of studying the characteristic patterns, reflected as percentage variations in some metrics, such as ranges, rms values, and harmonic distortion, that induction motors produce over vibration signals, electrical signals (stator current and fed voltages), and rotating speed.

Keywords: condition monitoring, induction motors, inverters

1. Introduction

Nowadays, induction motors are, and will remain in the future, the most important and frequently used electromechanical machines at industry facilities, but they are also of high interest in academic studies [1]. These machines are very important and are used because of their overall benefits, such as low cost, relatively easy manufacturing, robustness in performance, reliability, wide range of power capacities, and easy maintenance [2]. These motors represent around 80% of the used equipment and around 60% of the total energy consumption in the industry [3, 4]. Normally, electric motors are used with other elements such as mechanical couplings, drivers, and power sources to properly operate and to provide motion in the process [5].

The typical applications can be observed as domestic like in drills, pumps, blowers, vacuums, etc. or as industrial like in conveyors, machine tools, elevators, etc., which is the industrial application that is of main interest in this work [6]. However, the industrial environments are frequently changing and involve hard conditions related to electrical, mechanical, and thermal situations, among others, that directly or indirectly induce stresses to the induction motor [7]. These stresses cause malfunctioning, faults, or wear in many mechanical parts of the motor such as bearings, shafts, rotor bars, as well as in the electrical parts like in the stator and rotor windings [8]. The final consequences of these problems are observed as process downtimes during the motor operation and economic losses, which could be avoided through a correct monitoring and diagnosing process [9]. Thus, the importance of analyzing induction motors and their peripheral elements is justified because of their importance in the industry and domestic applications.

Regarding the importance of the induction motor, many works have been proposed with the purpose to develop methodologies capable of performing the monitoring and diagnosis of this industrial machine. For example, the work developed in [10] presents a methodology based on the motor current normalized residual harmonic analysis (MCNRHA) for diagnosing the rotor faults of broken bars and inter-turn short circuits in the stator windings. The residual harmonics are measured by means of the linear fast Fourier transform spectrum (LFFTS) of the healthy motor current signal and the faulty condition. In another example, the research presented in [11] calculates a fault intensity index in induction motors for the inter-turn short circuit fault of the stator winding. For this purpose, the raw current signal in the time domain obtained from the motor is processed through the discrete wavelet transform (DWT), and by using the detailed coefficients, the statistical parameter of the maximum norm is computed under several load conditions and fault severities. A common approach for detecting faults in the induction motor is based on the motor current signature analysis (MCSA); for instance, the work described in [12] presents an approach for detecting the misalignment fault. But that work uses the load torque signature analysis (LTSA) for studying a mechanism that transmits power between the motor and the loads by using different types of couplings. The obtained results show that these techniques perform according to the particular coupling defined. An interesting methodology for diagnosing the half-broken rotor bar (BRB) fault in an induction motor drive is developed in [13]. In that work, the motor was running under different operation conditions using a variable frequency drive (VFD), and the square of the current signal is analyzed because it generates more fault frequency components. To perform the diagnosis, the multiple signal classification (MUSIC) is implemented in an algorithm that can generate a pseudo-spectrum of the current signal. On one hand, the investigation developed by [14] presents an algorithm based on the Kalman filter (KF) for the stator inter-turn fault detection of induction motors. Thus, the KF is applied to extract the motor current signatures and motor voltage signatures; these signatures are later used for determining statistical fault indexes based on the standard deviation. A particular characteristic of the algorithm is that the effect of harmonic pollution is also analyzed, demonstrating to be effective in such conditions. As an alternative to the classical signal-based approaches such as voltages and currents, some other sensor signals have been considered for the monitoring and analysis of induction motors. For example, thermal image processing has been proposed, such as in [15], where three-phase induction motors are analyzed for fault detection. In this case, the thermal images are converted into a new color model for identification known as hue saturation and value (HSV), and then, five image segmentation methods are applied for obtaining the hue region: Sobel, Prewitt, Roberts, Canny, and Otsu. Next, different statistical parameters are obtained from the image matrices segmented for detecting three fault conditions under different load conditions: outer race bearing fault, inner race bearing fault, and ball bearing defects. On the other hand, the sound, acoustic, and vibration signals have also been addressed for analyzing induction motors and detecting faults. The case of acoustic signals is handled in [8], where the shortened method of multi-expanded frequency selection was developed together with the K-nearest neighbor (KNN) classifier. Meanwhile, the sound and vibration signals are adopted in the work reported in [16], by implementing the complete ensemble empirical mode decomposition (CEEMD) that divides the analyzed signals into intrinsic mode functions. Posteriorly, the frequency of the marginal Gabor representation is computed with the purpose of obtaining the spectral content in the frequency domain. The method was validated for two broken rotor bars. In summary, the several works discussed demonstrate that several methodologies through classical techniques have been reported, but not all the potential problems associated with induction motors have been addressed. For instance, many of the fault diagnostics are focused on the main elements of the motor, but other peripheral elements are not completely analyzed yet, such as those related to renewable energy systems.

In relation to renewable energy systems used for feeding through inverters and motor drives, some works have addressed this topic. An example of the renewable energy systems used for feeding induction motors is described in [17]. In such work, a system that supplies power energy through photovoltaic panels instead of a bank of batteries is presented. The proposed system integrates solar panels, a push-pull converter, and a pump (induction motor). The objective was to design a system by using the evaluation of the energy-processing cycles allowing optimizing a sensorless induction motor drive. In another case, a similar application was handled in the research reported by [18] that developed a simplified system for water pumping by using an induction motor and photovoltaic panels. This system considered two stages, the first being the extraction of the maximum power from the solar panel through the control of the duty ratio in a DC-DC boost converter using the maximum power point tracking (MPPT) technique. In the second stage, a source inverter operates the pump based on a scalar-controlled voltage way. However, the system was tested under different load conditions in a laboratory-controlled environment. On one hand, some works have focused on the motor feeding through the power inverter such as in [19]. That work has explored the topology of a power inverter type Z-source series for feeding an induction motor by using photovoltaic panels, considering that the inverter has a single-stage conversion with buck-boost capability. Additionally, from this study, it was concluded that the system has benefits such as current and harmonics reduction compared with a simple boost control scheme. In the same line, other works like the research presented by [20] developed a fuzzy logic controller for improving the speed response, reducing harmonic content, and enhancing the overall system performance of a multi-level power inverter used for feeding an induction motor. The proposed system integrates a photovoltaic panel, a boost converter, a multi-level inverter, a classical proportional-integral (PI) controller, and a three-phase induction machine; however, the performance of the approach is verified through simulations in MATLAB and Simulink. Finally, in [21], a power system based on auxiliary photovoltaic panels for electric vehicle applications is presented. The system novelty is a foldable scissor mechanism enabling the power system portability. Like the previous works, the validation of the system was carried out utilizing simulation experiments.

As observed from the previously discussed works, the effects of applying systems for supplying induction motors through devices based on renewable energy such as solar photovoltaic imply the use of power inverters. The effects of these inverters have been addressed from the system improvement viewpoint, considering, for example, hardware topologies. However, the effects of the power inverters considering the type of source output have not been completely analyzed yet and represent an area of opportunity.

In this chapter, the effects caused by the power inverters integrated in a renewable energy generation system, in islanding mode, over induction motors are analyzed. The analyzed system considers the connection of photovoltaic panels to two types of power inverters that are interchanged between the experimental tests. The connected inverter will supply an induction machine and a bank of batteries for power storing. Some physical magnitudes that consider electrical signals (current and voltage), vibration signals, and the motor speed will be acquired through a data acquisition system (DAS), for the behavior analysis. The power inverters considered are categorized into two main types: modified sine wave and pure sine wave. The main differences between the two devices are the internal hardware structures and topologies to generate the sine wave that will be used for supplying the motor. Additionally, the analysis of the inverters' effects will be done under different load conditions. Therefore, some metrics such as ranges, rms values, and harmonic distortion, obtained from the measured signals, will be presented and discussed, demonstrating the differences when the two types of power inverters feed the induction motor. Finally, the experimental results demonstrate how the motor operation varies depending on the type of power inverter by presenting noticeable variations in the percentage of the metrics when modified sine wave is used instead of pure sine wave.

2. Theoretical background

Next, the theoretical background of the power electronics used by solar photovoltaic generation systems (SPVGS), and even by some wind power generation systems (WPGS), will be addressed. The addressed theory focus particularly on the technologies that take the produced alternating current energy and convert it into direct current energy, better known as power inverters.

2.1 Power inverter

Nowadays, among all the types of renewable energy generation systems, solar PV technologies are the most frequently used for power generation because of the merits of the solar energy source, like its abundance (it is found practically in any place around the globe), less maintenance, no rotating or mechanical parts, low operational costs, and being pollution free. Thus, due to all these advantages, the SPVGS have more research and technological developments for all their integrated parts. In this sense, in the renewable energy generation systems, mainly in the SPVGS, it is very common to use power electronic components for providing an adequate energy output, for instance, through the power inverter. Essentially, the generation systems of renewable energy (GSRE) produce the output, in many cases, as direct current (DC) electricity, such as in the case of solar photovoltaic (SPV) systems. Additionally, this energy produced by the GSRE can be also stored in battery banks in the DC form due to several considerations, such as excess power generation, backup power, and limitations of the generation system. Such limitations in the energy generation process assume that the stored energy can be provided to the final user through the battery banks, as a stable DC power source, when the GSRE is unable to generate power energy, like at night for the case of SPVGS or lack of wind for the case of WPGS. Later, this DC output needs to be converted into alternating current (AC), which is because the loads used in domestic and business applications normally require the signal source in this AC form [22]. Therefore, a power inverter is a device that takes the generated power in the form of DC from the GSRE, or takes the energy stored in the battery bank, and turns it into AC power to operate the final loads [23]. Thus, the power inverter is a key device that normally comes together with the photovoltaic panels strictly used for obtaining an AC source.

It must be mentioned that due to the tendency of reduction in the prices for solar PV systems, it has caused an increment in the research about power inverters addressing considerations such as efficiency, size, weight, reliability. Therefore, today, the power inverter research industry has grown significantly and has developed a wide variety of inverter topologies with the purpose to meet the requirements of power conditioning. For example, Ref. [24] has presented a general classification of the existing inverters considering aspects such as the number of processing stages, the type of isolation, the power rating, the output shape, the voltage gain, the type of grid interface, and the soft/hard switching. Many of these topologies consider the signal output in the form of a pure sine wave, or a modified sine wave, with the main difference in the final applications. The topologies that use pure sine wave signal output are mainly designed for on-grid connections, meaning that the produced energy from the GSRE needs a signal capable of being synchronized and integrated into the commercial grid. In the other case, in many commercial, domestic, and a few cases of business applications where an off-grid connection will be used, better known as islanding appliances, a power inverter topology with modified sine wave signal output will be considered.

It is worth mentioning that although the power inverter is an important device, it can pollute the power signal, mainly affecting its quality by producing power disturbances (PD) such as spikes and harmonic content. This poor power quality (PQ) affects the final equipment fed by the power energy from the inverter and is reflected as losses and delivered heat, causing malfunctioning and even damage [24]. Therefore, the effects and impacts of the power inverters are a topic of interest because they are not fully addressed yet for application where induction motors are fed through renewable energy systems.

3. Experimental setup

The experimental test bench used for the validation of this proposal consists of a photovoltaic system that is used as a renewable energy generation system; the photovoltaic system is connected in an isolated way, and it is used to provide the power supply to a single-phase induction motor (IM); the general wiring of the aforementioned photovoltaic system is represented in **Figure 1**. As noted, the renewable energy generation system is composed of three solar cells (model SL150TU-18P), producing an average peak power of around 150 W per cell; a solar charge controller (model NV12V010E) is also taken into account since it is responsible for regulating the state-of-charge of a set of batteries (three batteries with model CL-31 T-700). Thus, the set of batteries allows to store the energy produced by the photovoltaic system

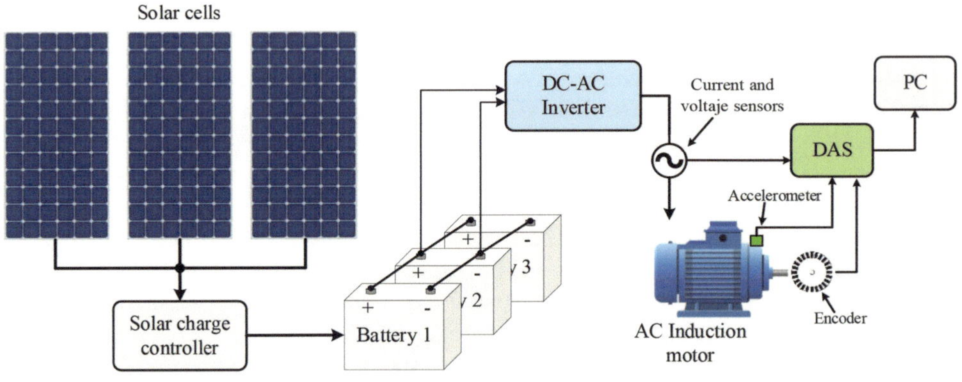

Figure 1.
General flowchart representing the wiring of the photovoltaic renewable energy generation system and the instrumentation to monitor several physical magnitudes.

(DC voltage), and then, the stored energy is converted into AC voltage by means of a DC-AC inverter. In this regard, it must be clarified that two different inverters are used for transforming the energy from DC to AC; one of the inverters is a 1500 W pure sine wave inverter (model RBP1500WRD by WZRELB), whereas the other one is a 1500 W modified sine wave inverter (INCO-1500 by TRUPER). The main technical characteristics of the two inverters are summarized in **Table 1**.

From **Table 1**, it is observed that both power inverters have similar characteristics; however, it is worth mentioning that the pure sine wave inverter is designed to limit the amount of harmonic components that can be found in the output signal. In this particular case, the THD value always remains under 5%, a situation that indicates that the resulting voltage signal is accomplished with international power quality standards. On the other hand, the modified sine wave inverter does not provide information regarding the THD. Nonetheless, this type of inverter is characterized by delivering a nearly squared signal; that is, the amount of harmonics is expected to be high. In fact, this last issue is one of the motivations for developing this study to find the repercussion of feeding an induction motor with a voltage signal that contains a high amount of harmonics and interharmonics.

Specification	Pure sine wave inverter	Modified sine wave inverter
Model	RBP1500WRD	INCO-1500
Manufacturer	WZRELB	TRUPER
Rated power	1500 W	1500 W
DC input voltage	10–15 V	10.5–15.5 V
AC output voltage	110 V	120 V
Frequency	60 Hz	60 Hz
Maximum efficiency	90%	85%
Total harmonic distortion (THD)	< 5%	Not provided by manufacturer

Table 1.
Technical characteristics for the DC-AC inverters used in the experimentation.

As stated, the use of these inverters is to analyze the effects that are produced over a 372 W single-phase IM (model 1RF20000DB004AB1 by SIEMENS) with two pairs of poles, efficiency around 62%, nominal rotating speed of 1755 RPM, and a nominal stator current consumption of 5.6 A (rms); the IM is used as the AC load connected to each one of the tested inverters.

Furthermore, for monitoring the operation of the IM, several sensors are installed to measure different physical magnitudes; therefore, a hall-effect current sensor (model SCT-013-030) is installed through the power supply lines to monitor the IM stator current consumption, and a transformer-based sensor (model ZMPT101B) is used to monitor the supply voltage; similarly, a triaxial accelerometer (model LIS3L02AS4) is installed over the IM case in the end-drive shaft in order to acquire the vibrations produced by the rotating movement, and an encoder is also installed in the IM shaft to measure the rotational speed. These sensors are installed and located as illustrated in **Figure 1**. The signals are acquired by means of two 12-bit 4-channel serial-output sampling analog-to-digital converters (ADS7841) that are mounted on a self-designed data acquisition system (DAS), which is based on a field programmable gate array (FPGA) technology. Thus, a sampling frequency of 6000 Hz is programmed in the proprietary DAS to acquire the stator current signal, the voltage signal, and the rotational speed, whereas the vibration signals are acquired with a sampling frequency of 3000 Hz. Accordingly, the aforementioned signals are continuously acquired during 30 seconds that comprise the start-up and the steady state of the IM, where the considered inverters are tested iteratively in order to perform several runs in the IM; all the acquired data are stored in a personal computer for posterior analysis.

4. Proposed methodology

The proposed methodology for analyzing the effects produced by pure sine and modified sine inverters in a single-phase IM consists of three main steps that are summarized in the flowchart of **Figure 2** that consist of *data acquisition*, *data processing*, and *analysis*.

In the *data acquisition* stage, the stator current, voltage, vibrations, and rotational speed are continuously acquired during 30 seconds of the IM operation; these signals are acquired by the proprietary DAS when the IM is fed through the pure sine and modified sine inverters, iteratively. In this regard, several tests are performed with each different inverter with the aim of comparing the repeatability of the experiments.

Subsequently, in the second stage of *data processing*, the acquired signals are subjected to a processing procedure by means of estimating the power spectral density

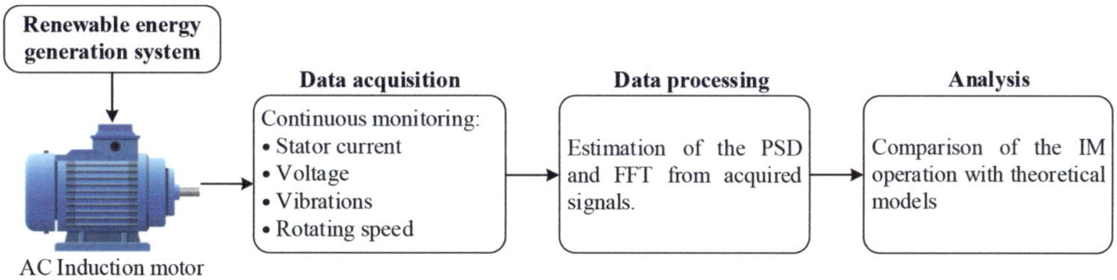

Figure 2.
Proposed methodology for analyzing the effects produced by pure sine and modified sine inverters in a single-phase IM.

(PSD) and by estimating the fast Fourier transform (FFT); precisely, the PSD is performed from the stator current, whereas the FFT is performed from the vibration signals. On the other hand, the raw voltage signal is analyzed in the time domain in order to compare similarities and differences between signals from the pure sine and modified sine inverters. Similarly, for both the inverters, the rotational speed signals are compared between them with an aim to identify the main differences, and such signals are also taken into account for a precise estimation of the theoretical frequencies that characterize the operation of the IM under study.

Finally, in the stage of *analysis*, the previous estimated PSD and FFT locate the theoretical frequencies that characterize the electrical and mechanical operation of the IM; that is, for the electrical signal, a frequency component must be located in f_s as the supply frequency component, and the mechanical operation is represented by f_r as the rotational speed; for both frequency components (f_s and f_r), some harmonics may appear according to the IM condition. In this sense, the most common faulty conditions that affect the operation of IMs are misalignment, unbalance, bearing defects, and broker rotor bars; thus, the harmonics of f_s and f_r may be more evident depending on whether the IM is working under the influence of any defect or not.

5. Results and discussions

The proposed method is performed in order to analyze the effects produced by pure sine and modified sine inverters in a single-phase IM; thus, the IM is fed through two different inverters, iteratively. Therefore, during the experimentation, several signals have been acquired and stored in a personal computer, and each measured signal comprises the start-up to the steady state of the IM; the signals are acquired during 30 seconds.

Accordingly, **Figure 3a** and **b** shows the rotational speed achieved by the IM in its end-drive shaft when the pure sine and modified sine inverters are used to feed such IM; the average speed that is reached during the steady state for each corresponding case is around 1799 rpm and 1794 ± 4 rpm, respectively. Thus, regarding the nominal speed provided by the manufactured IM, a speed variation between 2% and 3% is achieved when the IM is fed through both inverters. As shown in **Figure 3a** and **b**, there are some specific differences that characterize the working operation of the IM; probably, the main difference relies on the time that the IM requires to reach the steady state; precisely, **Figure 3a** depicts a soft start, where the IM achieves the steady state in an average time of around 1.9 seconds, whereas **Figure 3b** depicts an abrupt start, where the IM requires approximately 0.6 seconds to achieve its steady state. Thus, a soft start is preferred since it may lead to producing low stator currents during the start-up and may also benefit to avoid inducing structural damage to the whole elements that are linked to the IM, that is, rigid couplings, shafts, pulley belts, and gears, among others. Another difference that can be noted in **Figure 3a** and **b** is the stability of rotation when the IM has reached its steady state; that is, when the IM is fed by the pure sine inverter, a stable rotational speed is produced in the end-drive shaft (**Figure 3a**), whereas a variational rotating speed is generated when the modified sine inverter is used (**Figure 3b**). Thereby, most of the time, some processes and/or applications are affected when the rotational speed of the IM that drives such processes is variable; that is, an electric water pump may produce a variable flow of water if the rotating speed of impellers is not constant. Moreover, the sudden occurrence

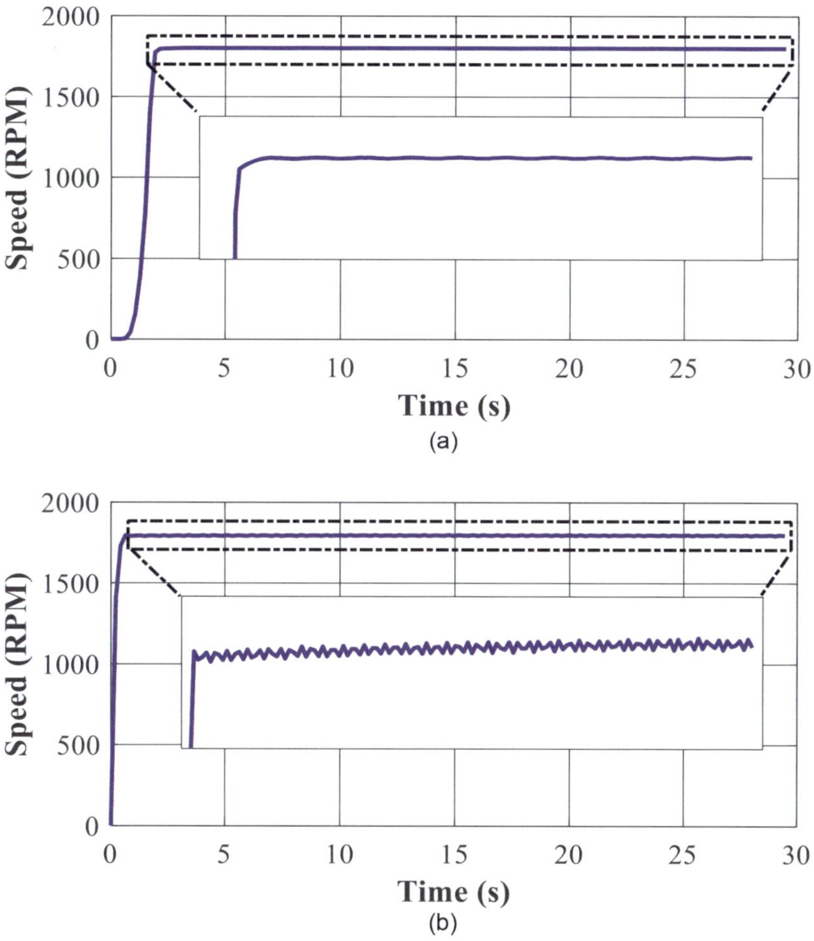

Figure 3.
Rotational speed achieved in end-drive shaft of the IM when it is fed through a) the pure sine inverter and b) the modified sine inverter.

of vibrations may affect and induce damage over the whole elements linked to the IM that experiences variable speed.

Consequently, **Figure 4a** and **b** shows the measurements of the stator current consumption when the IM was fed through the pure sine and modified sine inverters, respectively. As it can be appreciated, the soft starting produced by the pure sine inverter also leads to a reduced current consumption during the start-up; such current consumption is approximately 11.5 amperes peak. Once the IM has achieved its steady state, the average current consumption is around 6.5 amperes peak, and although **Figure 4a** depicts the stator current in the IM when it is fed with the pure sine inverter, an almost perfect sine wave with some speaks is measured. In this sense, the current signal is dirty since the AC-DC inverter is composed of several power electronic elements. On the other hand, high current consumption is demanded by the IM when it is fed with the modified sine inverter as illustrated in **Figure 4b**, where a peak current consumption higher than 25 amperes is reached and an average current consumption of around 9.4 amperes is produced in the steady state. In addition, the shape of the current wave in **Figure 4b** is not sinusoidal and consumption

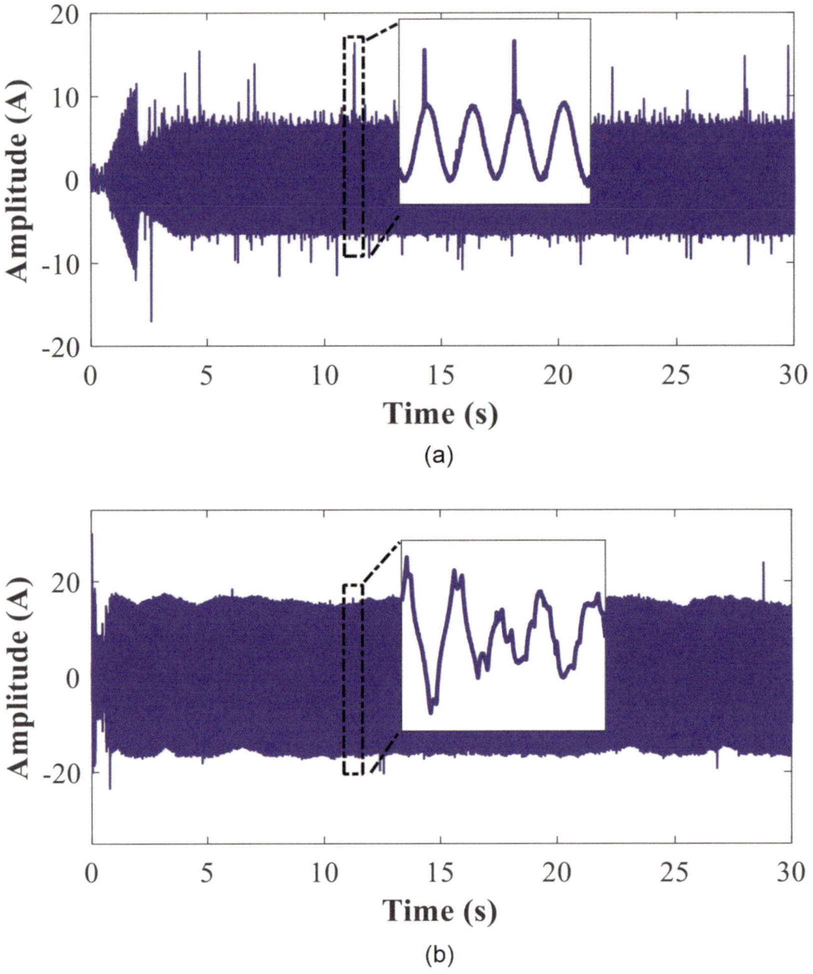

Figure 4.
Measurements of the stator current consumption produced in the IM when it is fed through a) the pure sine inverter and b) the modified sine inverter.

shows several variations; this fact is due to the power quality characteristics that have the modified sine inverter. Thus, the most critical increase in the stator current consumption is presented during the start-up where the increase is more than 200% by comparing the current peak of **Figure 4a** and **b**; meanwhile, percentages around 82% and 118% are achieved during the steady state in comparison with the nominal current consumption given by the manufacturer.

Voltage signals are shown in **Figure 5a** and **b**, respectively, for each one of the considered inverters, pure sine and modified sine. As it is observed, both voltage signals are generated with a pike amplitude of about 180 volts; also, both signals show distortion that in terms of power quality can be understood as harmonics, sags, swells, and transients, among others. Precisely, in the zoomed-in view shown in **Figure 5a**, a transient is noted, which is also present in the whole voltage signal; on the other hand, the zoomed-in view of **Figure 5b** also shows transients that in general may affect the proper operation of the IM. Hence, all the electronic elements included in both inverters are the main source that leads to producing voltage signals with high distortion.

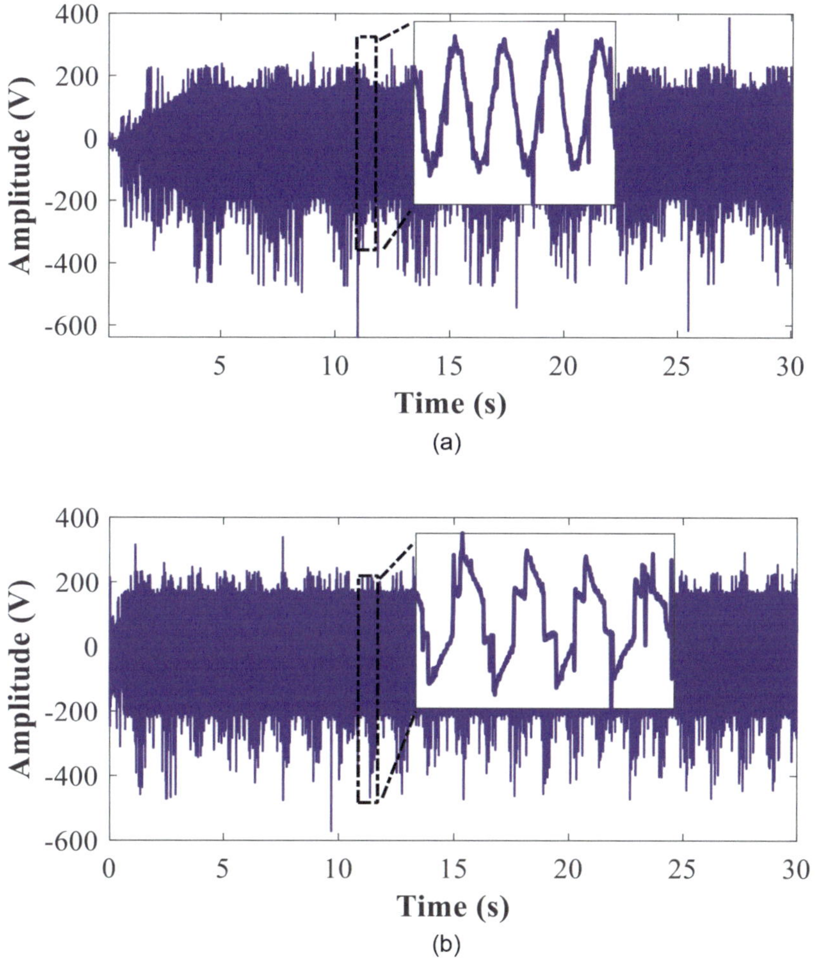

Figure 5.
Recorder voltage signals produced by the considered AC-DC inverters to feed the IM: a) the pure sine inverter and b) the modified sine inverter.

On the other hand, the main difference between the voltage signals of **Figure 5a** and **b** is the shape, where the voltage signal in **Figure 5a** depicts an almost perfect sine wave, while the voltage signal in **Figure 5b** does not have a sinusoidal wave. In this regard, it must be highlighted that a high harmonic content is inherent in the voltage generation for both inverters. Additionally, the THD is estimated from both acquired signals in order to validate the percentages of distortion given by the manufactures; in this way, distortions of around 5.5% and 22.3% are computed for both inverters, pure sine and modified sine, respectively.

The acquired vibration signals are associated with the mechanical operation of the IM; in fact, any AC rotating machine may experiment with different operating conditions when fed through a renewable power supply by means of inverters. For example, if an AC electric motor is connected to a power supply, the power quality can affect the operation of the device, producing speed variations since it is in function of the voltage applied. Subsequently, the acquired vibrations when the IM is fed through the pure and modified sine inverters are shown in **Figures 6a** and **b**, respectively; as it is observed, the vibrations produced in the

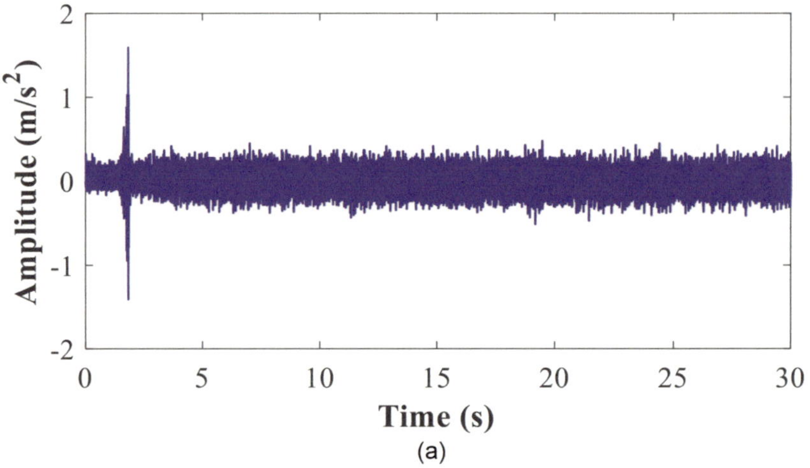

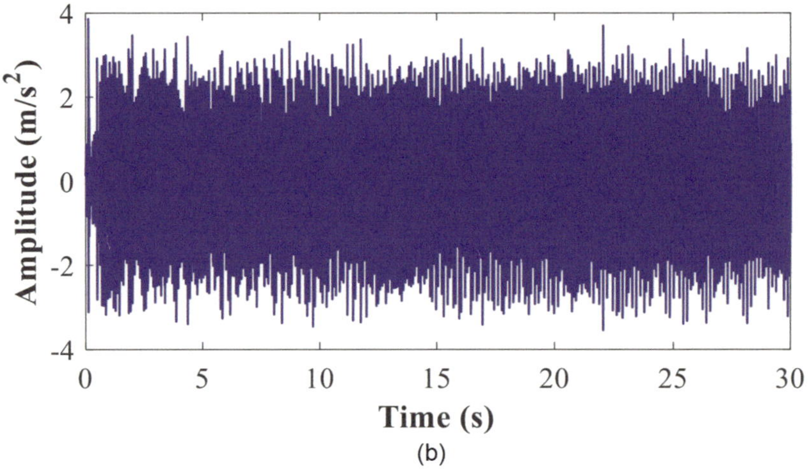

Figure 6.
Acquired vibration signal in the radial axis of the IM when it is fed through both inverters: a) the pure sine inverter and b) the modified sine inverter, respectively.

IM have higher amplitude when it is fed through the modified sine inverter, while the level of vibrations is reduced when the pure sine inverter is used as the power supply source. Also, there are some differences between **Figure 6a** and **b**; that is, a transient vibration spike is presented at the end of the start-up of the IM when it is fed with the pure sine inverter, and then, the vibration level is retained with a specific average amplitude. On the other hand, the IM experiences a high level of vibrations, which apparently is retained during the start-up and the steady state of the IM when the modified sine inverter is used. Qualitatively, by comparing the RMS values of both vibration signals, an increase from 0.0997 to 0.6989 is presented when the IM is fed with sine and modified inverters. For the IM under test, the vibrations are directly produced by the quality of the voltage signal that is provided by both the considered inverters; it should be mentioned that the IM is in a healthy condition; thus, there are no external factors that lead to introducing the occurrence of vibrations. **Table 2** summarizes the most important aspects depicted with the use of pure and modified sine inverters.

Physical magnitude	Pure sine wave inverter	Modified sine wave inverter
Speed	The speed is constant and reaches an average value of around 1799 RPM	The speed shows fluctuations producing an average speed value of 1794 ± 4 RPM
Current	The stator current consumption is less than the nominal current consumption	An increase of around 118% of the stator current consumption is reached during the steady state
Voltage	The distortion is around 5.5% as manufacturer depicts	The distortion is higher than 20% as the shape wave does not match the sine wave
Vibration	An averaged RMS vibration value of 0.0997 is achieved with a soft noise	The vibration increases seven times by considering the RMS vibration value 0.6989

Table 2.
Technical characteristics for the DC-AC inverters used in the experimentation.

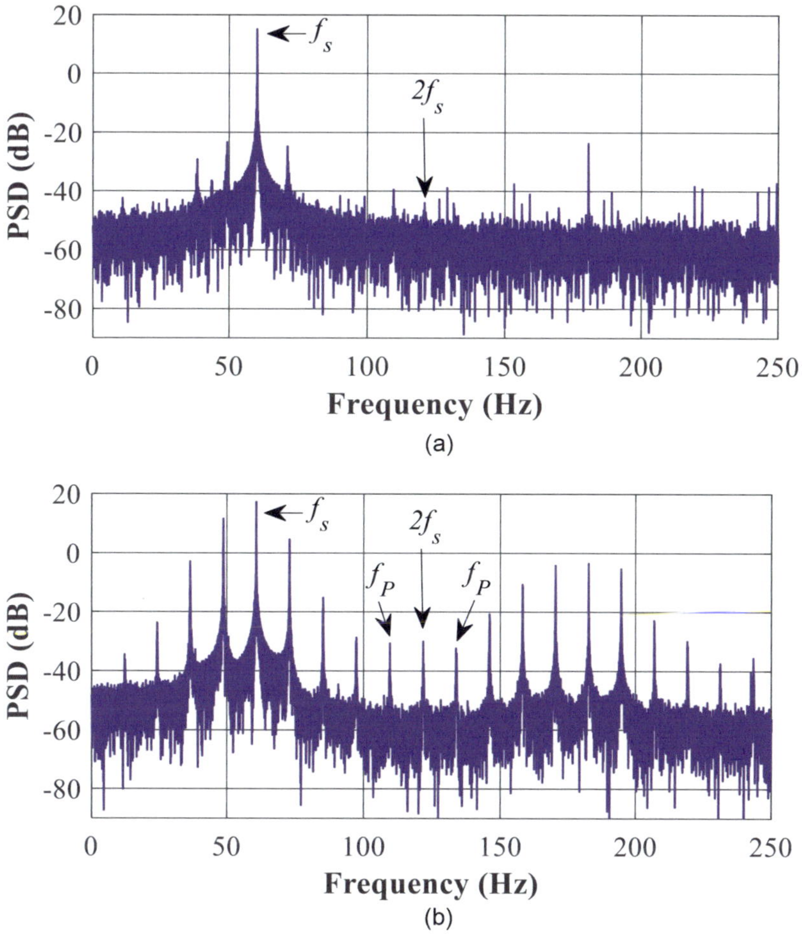

Figure 7.
Achieved PSD of the acquired stator current when the IM is fed through a) the pure sine inverter and b) the modified sine inverter, respectively.

Accordingly, the PSD of the stator current and the FFT of the vibrations are carried out in order to provide a more accurate analysis of the effects that pure sine and modified sine inverters introduce over the operation of an IM. Therefore, **Figure 7a** and **b** shows the obtained PSD when the pure and modified inverters are used to feed the IM, respectively. Some aspects must be highlighted from these PSDs; the first one is that the supply frequency (f_s) is present and it may be located approximately around 60 Hz, and it appears with a high amplitude; the second one is that in the PSD of **Figure 7b** appears a significant number of frequency components with higher amplitude in comparison with the PSD of **Figure 7a**. The appearance of additional frequency components is also due to the quality of the power voltage supply; moreover, as expected, the use of modified sine inverters results in the introduction of frequency components over the PSD that may be masked and/or confused with those characteristic fault-related frequency components that are commonly induced by faults such as misalignments, unbalances, short circuits, and broken rotor bars, among others. Specifically, in the PSD of **Figure 7b** are induced several harmonics related to f_s ; such harmonics appear at $2f_s$ and $3f_s$, but the $2f_s$

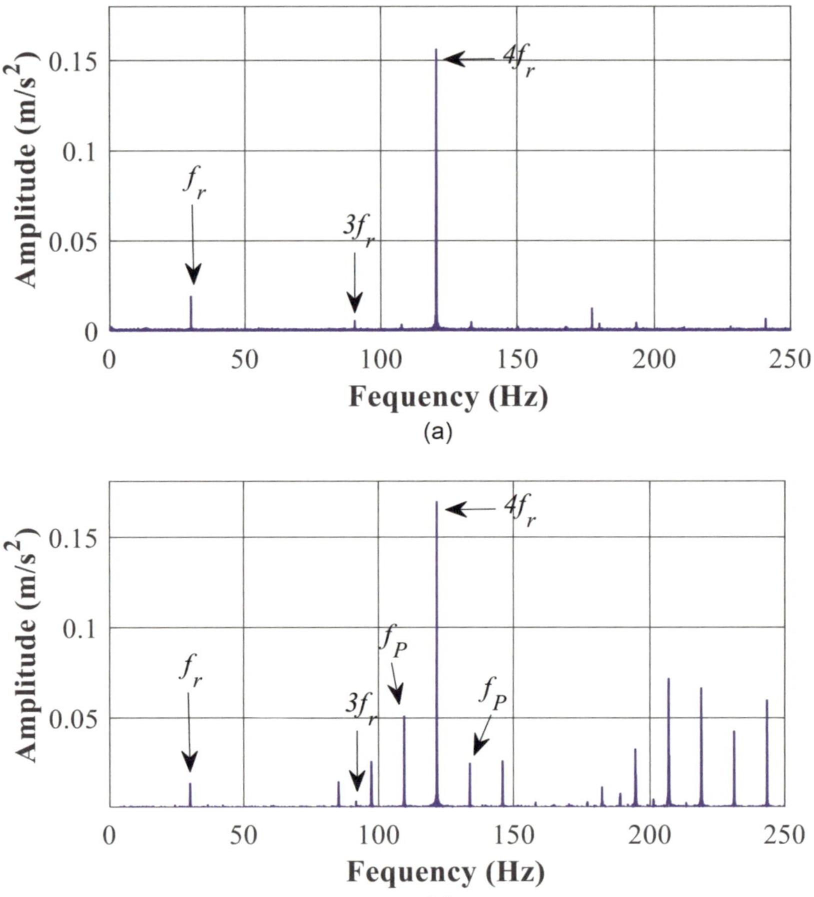

Figure 8.
Achieved FFT of the acquired vibrations when the IM is fed through a) the pure sine inverter and b) the modified sine inverter, respectively.

components may increase their amplitude when problems associated with phase problems are inherent to the power supply; additionally, sidebands may appear around $2f_s$ that are associated with the pole pass frequency (f_P). Thereby, in **Figure 7b**, a significant increase of frequency components ($2f_s$ and $3f_s$) as well as the occurrence of additional frequency components (f_P) that ideally depict malfunction problems in the power supply is observed.

Finally, the vibration signals are processed by means of estimating the frequency spectra through the FFT; thus, **Figure 8a** and **b** shows the resulting vibration spectra that belong to the IM operation when it is fed with the pure and modified sine inverters, respectively. As it is observed in both spectra, some characteristics related to frequency components appear, where it is important to identify the frequency component associated with rotational speed (f_r) that is reached in the IM; also, there are some harmonics of f_r that can be identified over both spectra such as the third and fourth harmonics ($3f_r$ and $4f_r$). The aforementioned harmonics are the most important since it is around these frequency components that sideband frequencies appear, separated by the pole pass frequency (f_P) when the IM is fed through the modified sine inverter. Thus, the quality of the voltage power supply also affects the mechanical operation of electric rotating machines such as IM even if such machines are in healthy condition; such affectations commonly produce variations in the rotational speed, generation of abnormal noise during the steady state, and sometimes the occurrence of structural vibrations that affect the whole components of the IM. On the other hand, the use of pure sine inverters allows to operate the IM almost as it was working under conventional conditions where the power supply voltage is provided by an electrical factory.

6. Conclusions

This work presented an analysis of the effects produced in an IM when it is fed through renewable energy by means of pure sine and modified sine inverters; the analysis consists of acquiring some physical magnitudes such as the stator current that is consumed by the IM, the voltage supplied to the IM, the rotational speed reached in the end-drive shaft of the IM, and the mechanical vibrations produced by the IM during its working operation. These signals are continuously acquired and are first compared in the time domain. Regarding the obtained results, it can be concluded that a slight reduction in the stator current consumption is achieved when the IM is fed through the pure sine inverter; also, a soft start-up is produced, the average rotational speed is retained in the end-drive shaft, as well as vibration levels are kept low. Meanwhile, when the IM is fed through the modified inverter, the stator current consumption increases, and this increase may lead to an increase in the temperature of the IM and several damages can be also produced. Additionally, the modified inverter makes the IM rotate with variations and noise, and subsequently, the occurrences of vibrations are present in the whole IM case. Finally, the stator current and vibrations are also analyzed by means of the PSD and FFT, where significant differences are appreciated by comparing the resulting spectra when the IM is fed with both the considered inverters. The most important aspect to be highlighted from the PSD and FFT is that the modified sine inverter leads to the introduction of a frequency component that is associated with the pole pass frequency, where such a component is associated with phase problems. Finally, pure sine inverters are recommended in applications where rotating machines are involved; in fact, the consideration of pure sine inverters may result in extending the useful life of those elements that are fed through them as much as possible.

Acknowledgements

This work has been partially supported by the "FONDO PARA EL DESARROLLO DEL CONOCIMIENTO" (FONDEC-UAQ 2022) under the project "Analysis and detection of failures in alternative energy power generation systems through artificial intelligence algorithms and machine learning techniques".

Author details

Arturo Yosimar Jaen-Cuellar, David Alejandro Elvira-Ortiz,
Emmanuel Resendiz-Ochoa and Juan Jose Saucedo-Dorantes*
Engineering Faculty, Autonomous University of Queretaro, San Juan del Rio, Mexico

*Address all correspondence to: juan.saucedo@uaq.mx

References

[1] Zamudio-Ramirez I, Osornio-Rios RA, Antonino-Daviu JA, Razik H. Magnetic flux analysis for the condition monitoring of electric machines: A review. IEEE Transactions on Industrial Informatics. 2022;**18**(5):2895-2908

[2] Corral-Hernandez JA, Antonino-Daviu JA. Thorough validation of a rotor fault diagnosis methodology in laboratory and field soft-started induction motors. Chinese Journal of Electrical Engineering. 2018;**4**(3):66-72

[3] Kumar S, Mukherjee D, Guchhait PK, Banerjee R, Srivastava AK, Vishwakarma DN, et al. A comprehensive review of condition based prognostic maintenance (CBPM) for induction motor. IEEE Access. 2019;7:90690-90704

[4] Errigo A, Choi JK, Kissock K. Techno-economic-environmental impacts of industrial energy assessment: Sustainable industrial motor systems of small and medium-sized enterprises. Sustainable Energy Technologies and Assessments. 2022;**49**:101694

[5] Vadi S, Bayindir R, Toplar Y, Colak I. Induction motor control system with a Programmable Logic Controller (PLC) and Profibus communication for industrial plants — An experimental setup. ISA Transactions. 2022;**122**:459-471

[6] Zhang J, Wang P, Gao RX, Sun C, Yan R. Induction motor condition monitoring for sustainable manufacturing. Procedia Manufacturing. 2019;**33**:802-809

[7] Jeffali F, Ouariach A, El Kihel B, Nougaoui A. Diagnosis of three-phase induction motor and the impact on the kinematic chain using non-destructive technique of infrared thermography. Infrared Physics & Technology. 2019;**102**:102970

[8] Glowacz A. Fault diagnosis of single-phase induction motor based on acoustic signals. Mechanical Systems and Signal Processing. 2019;**117**:65-80

[9] Vishwakarma M, Purohit R, Harshlata V, Rajput P. Vibration analysis & condition monitoring for rotating machines: A review. Materials Today: Proceedings. 2017;**4**:2659-2664

[10] Allal A, Khechekhouche A. Diagnosis of induction motor faults using the motor current normalized residual harmonic analysis method. International Journal of Electrical Power & Energy Systems. 2022;**141**:108219

[11] Almounajjed A, Sahoo AK, Kumar MK. Diagnosis of stator fault severity in induction motor based on discrete wavelet analysis. Measurement. 2021;**182**:109780

[12] Verucchi C, Bossio J, Bossio G, Acosta G. Misalignment detection in induction motors with flexible coupling by means of estimated torque analysis and MCSA. Mechanical Systems and Signal Processing. 2016;**80**:570-581

[13] Singh G, Naikan VNA. Detection of half broken rotor bar fault in VFD driven induction motor drive using motor square current MUSIC analysis. Mechanical Systems and Signal Processing. 2018;**110**:333-348

[14] Namdar A, Samet H, Allahbakhshi M, Tajdinian M, Ghanbari T. A robust stator inter-turn fault detection in induction motor utilizing Kalman filter-based algorithm. Measurement. 2022;**187**:110181

[15] Al-Musawi AK, Anayi F, Packianather M. Three-phase induction motor fault detection based on thermal image segmentation. Infrared Physics & Technology. 2020;**104**:103140

[16] Delgado-Arredondo PA, Morinigo-Sotelo D, Osornio-Rios RA, Avina-Cervantes JG, Rostro-Gonzalez H, Romero-Troncoso R. Methodology for fault detection in induction motors via sound and vibration signals. Mechanical Systems and Signal Processing. 2017;**83**:568-589

[17] Vitorino MA, de Rossiter Correa MB, Jacobina CB, Lima AMN. An effective induction motor control for photovoltaic pumping. IEEE Transactions on Industrial Electronics. 2011;**58**(4):1162-1170

[18] Singh B, Sharma U, Kumar S. Standalone photovoltaic water pumping system using induction motor drive with reduced sensors. IEEE Transactions on Industry Applications. 2018;**54**(4):3645-3655

[19] Adle R, Renge M, Muley S, Shobhane P. Photovoltaic based series z-source inverter fed induction motor drive with improved shoot through technique. Energy Procedia. 2017;**117**:329-335

[20] Chandrasekaran S, Durairaj S, Padmavathi S. A Performance evaluation of a fuzzy logic controller-based Photovoltaic-fed multi-level inverter for a three-phase induction motor. Journal of the Franklin Institute. 2021;**358**:7394-7412

[21] Jin Z, Li D, Hao D, Zhang Z, Guo L, Wu X, et al. A portable, auxiliary photovoltaic power system for electric vehicles based on a foldable scissors mechanism. Energy and Built Environment. 2022;**2022**:10

[22] Gorjian S, Shukla A, editores. Photovoltaic Solar Energy Conversion: Technologies, Applications and Environmental Impacts. 1st ed. Cambridge: Elsevier; 2020

[23] Mayfield R. Photovoltaic Design & Installation for Dummies. Hoboken, NJ: Wiley; 2010. p. 366

[24] Dogga R, Pathak MK. Recent trends in solar PV inverter topologies. Solar Energy. 2019;**183**:57-73

Section 2

Electric Motor Control

Chapter 4

Robust Mechanism for Speed and Position Observers of Electrical Machines

Marcin Morawiec

Abstract

In the sensorless control system, the rotor speed or position is not measured but reconstructed in the dedicated observer structure. The observer structure is based on the mathematical model of an electrical machine. This model is often determined in the space vector form by using the stator/rotor flux vector and stator/rotor current vector components. During the machine works, there exist working points in which the observer can be unstable or its accuracy is unsatisfactory. In order to increase the observer system stability, the Lyapunov theorem should be satisfied. Using this, the observer system's proper stabilizing function can be determined. However, in many cases, this procedure is not sufficient and in close to an unstable region properties of the speed observer structure are very poor—the estimation errors have values exceeding 5%, which causes loss of synchronization in case of synchronous machines and errors in the values of electromagnetic torque or stator/rotor fluxes. In order to prevent this undesirable phenomenon, additional laws of estimation should be introduced to the speed or position estimation mechanism, which is proposed in this chapter. This mechanism is named in this chapter "robust" because during the machine works, it increases significantly the properties of the whole sensorless control system, minimizing the speed or position estimation errors almost to zero, close to the unstable region (small rotor speed with the regenerating machine mode or close to synchronous of rotor speed in case of the doubly fed generator). The proposed robust mechanism has been tested by using simulation and experimental investigations prepared for: the squirrel-cage induction machine, permanent magnet synchronous machine, and doubly fed induction generator.

Keywords: speed estimation, rotor position, adaptive, non-adaptive, induction machine, permanent magnet synchronous machine, doubly fed induction generator

1. Introduction

In sensorless control of an electrical machine, the rotor speed value or rotor position is not measured but reconstructed by an observer structure. In the literature, methods of reproducing the rotor speed or rotor position can be divided into three [1]: algorithmic, neural network, and physical methods. The most popular is an algorithmic method in which the observer structure is based on the mathematical model of an electrical machine. This group includes state full and reduced-order observers, [2], the adaptive full-order observer (AFO), [3], Kalman filters, [4], model reference adaptive observers MRAS, [5], sliding mode observers, [6], and backstepping, [6]. The other approach to the estimation of the state variables is to extend the model of a machine with an additional state variable—an auxiliary state, [7]. The rotor speed value in these observers can be reconstructed from the classical adaptation law by using the proportional-integral controller (PI), [1, 7, 8]. The rotor position value can be obtained by using the integration of the rotor speed value in the same integration step, [7, 8]. Other approach to the reconstruction of the rotor speed value is the non-adaptive method. The rotor speed value is obtained by using the suitable algebraic transformation of the estimated state variables, [5, 6].

The main problem in the sensorless control systems is the stability of the observer structure in the wide changes of working points of the machine, [8, 9]: from zero to nominal rotor speed, under load torque injections, and for regenerating mode. Stabilization of the observer structure under regenerating mode and low speed of the induction machine, IM, was studied in many papers, [9–11]. For this case, the frequency of stator voltage is almost zero, and there exist unstable poles of the observer system, [8, 10]. Similarly, the problem occurs for the permanent or interior permanent magnet synchronous machines (PMSM/IPMSM) during the zero rotor speed; while the electromagnetic force (EMF) is not generated, [12–14]. To overcome this problem, a different value of stator current or voltage (high [13] or low [14] frequency) is injected into the stator voltage from an inverter.

A robust mechanism for the rotor speed estimation is proposed in this chapter. The proposed approach is suitable for the speed observer structures, which are based on a mathematical model of an electrical machine (algorithmic) in the space vector form. In Section 2, the mathematical model of an electrical machine is considered in the general form for the nonlinear class of systems. In Section 3, the application to IM is shown. In Section 4, the speed observer of IPMSM with the robust mechanism is proposed. In Section 5, the robust mechanism for the rotor speed and position estimation is adapted to the observer structure of DFIG.

All the theoretical derivations are confirmed by using simulation and experimental investigations.

2. Design procedure of the speed and position observer

One of the most popular design procedures for the speed observer of an electrical machine is based on the second theorem of the Lyapunov of asymptotical stability of the system in the general form

$$\dot{x} = Ax + Bu, \tag{1}$$

$$y = Cx, \tag{2}$$

where is assumed that $\boldsymbol{A}$, $\boldsymbol{B}$, and $\boldsymbol{C}$ are the matrixes that including the system parameters, $\boldsymbol{x}$ is the vector of state variables, and $\boldsymbol{u}$ is the vector of controls.

Considering only (1) the model can be rewritten to the vector components form in which (αβ) is the stationary reference frame and the index k means the number of the state variables in the system defined in (1)–(2)

$$\dot{x}_{k\alpha} = a_k x_{k\alpha} + \dots + b_k u_{k\alpha}, \tag{3}$$

$$\dot{x}_{k\beta} = a_k x_{k\beta} + \dots + b_k u_{k\beta}. \tag{4}$$

If the system (3)–(4) will be connected to the rotating reference frame, then the differential equations have the form

$$\dot{x}_{kd} = a_k x_{kd} + \omega_{dq} x_{kq} + \dots + b_k u_{kd}, \tag{5}$$

$$\dot{x}_{kq} = a_k x_{kq} - \omega_{dq} x_{kd} + \dots + b_k u_{kq}, \tag{6}$$

where ω_{dq} is the angular speed of the (d-q) reference frame, and (u_{kd}, u_{kq}) are the controls defined in (d-q).

It can be assumed that the system model belongs to the operation domain ***D*** defined by the set of values

$$\mathrm{D} = \left\{ \mathrm{x} \in \mathrm{R}^k,\ |x_{kd}| \le x_{kd}^{\max},\ |x_{kq}| \le x_{kq}^{\max},\ \omega_{dq} \le \omega_{dq}^{\max} \right\}, \tag{7}$$

where

$x_{kd}^{\max}, x_{kq}^{\max}, \omega_{dq}^{\max}$ are the maximum values for the state variables, and the parameters in the system a_k, b_k have known, constant, and bounded values.

Assumption 1. For the system (5)–(6) in which the ω_{dq} is treated as the parameter, it is possible to reconstruct its value by using the adaptive and non-adaptive approaches and state of variables x_k. Moreover, the controls (u_{kd}, u_{kq}) satisfied the persistent of excitation condition [14].

The first step in the procedure of design of the observer structure is to stabilize the observer for the system (5)–(6). The observer structure has the following form:

$$\dot{\hat{x}}_{kd} = a_k \hat{x}_{kd} + \hat{\omega}_{dq} \hat{x}_{kq} + \dots + b_k u_{kd} + v_d, \tag{8}$$

$$\dot{\hat{x}}_{kq} = a_k \hat{x}_{kq} - \hat{\omega}_{dq} \hat{x}_{kd} + \dots + b_k u_{kq} + v_q, \tag{9}$$

where the estimated values are marked by "^", and v_d, v_q are the inputs to the observer (8)–(9), which stabilize the system.

The estimation errors between estimated (8)–(9) and real/measured values (5)–(6) are expressed by

$$\tilde{x}_{kd,q} = \hat{x}_{kd,q} - x_{kd,q}, \tag{10}$$

$$\tilde{\omega}_{dq} = \hat{\omega}_{dq} - \omega_{dq}. \tag{11}$$

For the above-defined estimation errors, it is possible to determine the model of estimation errors, which form is as follows:

$$\dot{\tilde{x}}_{kd} = a_k \tilde{x}_{kd} + \hat{\omega}_{dq} \tilde{x}_{kq} + \tilde{\omega}_{dq} \hat{x}_{kq} - \tilde{\omega}_{dq} \tilde{x}_{kq} + v_d, \tag{12}$$

$$\dot{\tilde{x}}_{kq} = a_k \tilde{x}_{kq} - \hat{\omega}_{dq} \tilde{x}_{kd} - \tilde{\omega}_{dq} \hat{x}_{kd} + \tilde{\omega}_{dq} \tilde{x}_{kd} + v_q. \tag{13}$$

The next step is to determine the form of stabilizing functions, which stabilize the observer structure (8)–(9). By using the Lyapunov theorem, the observer structure will be stable if the candidate of the Lyapunov function

$$V = 0.5\left(\tilde{x}_{kd}^2 + \tilde{x}_{kq}^2\right) + V_1 \geq 0, \tag{14}$$

is positively defined, and $V_1 > 0$ has the form

$$V_1 = \gamma^{-1}\tilde{\omega}_{dq}^2. \tag{15}$$

Derivative of Lyapunov function (14) must be negative determined, therefore using (12)–(13), its form is determined

$$\begin{aligned}\dot{V} &= \tilde{x}_{kd}\left(a_k\tilde{x}_{kd} + \hat{\omega}_{dq}\tilde{x}_{kq} + v_d\right) + \tilde{x}_{kq}\left(a_k\tilde{x}_{kq} - \hat{\omega}_{dq}\tilde{x}_{kd} + v_q\right) \\ &\quad + \tilde{\omega}_{dq}\left(\gamma^{-1}\dot{\tilde{\omega}}_{dq} + \hat{x}_{kq}\tilde{x}_{kd} - \hat{x}_{kd}\tilde{x}_{kq}\right) \leq 0.\end{aligned} \tag{16}$$

The observer structure will be asymptotically stable if the Lyapunov theorem is satisfied and the stabilizing functions are chosen

$$v_d = -a_k\tilde{x}_{kd}, \tag{17}$$

$$v_q = -a_k\tilde{x}_{kq}, \tag{18}$$

then derivative (16) has the form

$$\dot{V} = \tilde{\omega}_{dq}\left(\hat{x}_{kq}\tilde{x}_{kd} - \hat{x}_{kd}\tilde{x}_{kq}\right) \leq 0. \tag{19}$$

To satisfy (19), the value of the parameter $\tilde{\omega}_{dq}$ should be determined by

$$\dot{\tilde{\omega}}_{dq} = -\gamma\left(\hat{x}_{kq}\tilde{x}_{kd} - \hat{x}_{kd}\tilde{x}_{kq}\right), \tag{20}$$

where

$\gamma > 0$ is the tuning gain.

For (20), the derivative of the Lyapunov function is always smaller than zero $\dot{V} < 0$, and the Lyapunov condition is satisfied.

2.1 Adaptive estimation of parameter ω_{dq}

The estimated value of the parameter $\hat{\omega}_{dq}$ can be determined from (20) under assumption that the derivative of the real value is constant in time $\dot{\omega}_{dq}$ and equal to zero

$$\dot{\hat{\omega}}_{dq} = -\gamma\left(\hat{x}_{kq}\tilde{x}_{kd} - \hat{x}_{kd}\tilde{x}_{kq}\right). \tag{21}$$

The above estimation law is named in the literature [15] as the classical adaptation law.

Remark 1. The assumption $\dot{\omega}_{dq} = 0$ is not desirable for the nonlinear system, in which the highest accuracy of estimation is needed. For $\dot{\omega}_{dq} \neq 0$ and $\omega_{dq} = \hat{\omega}_{dq} - \tilde{\omega}_{dq}$, after substitution (17)–(18) to (16), the derivative of the Lyapunov function has the following form:

$$\dot{V} = \tilde{\omega}_{dq}\left(-\frac{1}{\gamma}\left(\dot{\hat{\omega}}_{dq} - \dot{\tilde{\omega}}_{dq}\right)\right) \leq 0. \tag{22}$$

After substitution (21) to (22), the update form of derivative of the Lyapunov function is achieved

$$\dot{V} = -\tilde{\omega}_{dq}\left(-\left(\hat{x}_{kq}\tilde{x}_{kd} - \hat{x}_{kd}\tilde{x}_{kq}\right) + \frac{1}{\gamma}\dot{\tilde{\omega}}_{dq}\right) \leq 0. \tag{23}$$

It is easy to check that in (23) the dependence in the internal bracket $(\hat{\mathbf{x}}_k \times \tilde{\mathbf{x}}_k) = \hat{x}_{kq}\tilde{x}_{kd} - \hat{x}_{kd}\tilde{x}_{kq}$ means the cross-product of the pair of two vectors that occur in the observer system. The cross-product for can be determined by using Lagrange's identity [16].

Assumption 2. Considering the pair of vectors $(\hat{\mathbf{x}}_k, \tilde{\mathbf{x}}_k)$ defined in the observer system (8)–(9) and for the estimation errors (10)–(13), there exists Lagrange's identity [16], which has the following form:

$$(\hat{\mathbf{x}}_k \times \tilde{\mathbf{x}}_k)^2 \equiv |\hat{\mathbf{x}}_k|^2|\tilde{\mathbf{x}}_k|^2 - (\hat{\mathbf{x}}_k \cdot \tilde{\mathbf{x}}_k)^2. \tag{24}$$

Considering the vector components defined in (d-q) reference frame (24) can be rewritten as

$$\tilde{x}_{kd}\hat{x}_{kq} - \tilde{x}_{kq}\hat{x}_{kd} = \sqrt{\left(\hat{x}_{kd}^2 + \hat{x}_{kq}^2\right)\left(\tilde{x}_{kd}^2 + \tilde{x}_{kq}^2\right) - \left(\hat{x}_{kd}\tilde{x}_{kd} + \hat{x}_{kq}\tilde{x}_{kq}\right)^2}. \tag{25}$$

Substituting (25) to (23), the derivative of the Lyapunov function has the following form:

$$\dot{V} = \tilde{\omega}_{dq}\left(\sqrt{\left(\hat{x}_{kd}^2 + \hat{x}_{kq}^2\right)\left(\tilde{x}_{kd}^2 + \tilde{x}_{kq}^2\right) - \left(\hat{x}_{kd}\tilde{x}_{kd} + \hat{x}_{kq}\tilde{x}_{kq}\right)^2} - \frac{1}{\gamma}\dot{\tilde{\omega}}_{dq}\right) \leq 0. \tag{26}$$

The Lyapunov theorem is satisfied if

$$\dot{\tilde{\omega}}_{dq} = \gamma\sqrt{\left(\hat{x}_{kd}^2 + \hat{x}_{kq}^2\right)\left(\tilde{x}_{kd}^2 + \tilde{x}_{kq}^2\right) - \left(\hat{x}_{kd}\tilde{x}_{kd} + \hat{x}_{kq}\tilde{x}_{kq}\right)^2}, \tag{27}$$

where $\left(\left(\hat{x}_{kd}^2 + \hat{x}_{kq}^2\right)\left(\tilde{x}_{kd}^2 + \tilde{x}_{kq}^2\right) - \left(\hat{x}_{kd}\tilde{x}_{kd} + \hat{x}_{kq}\tilde{x}_{kq}\right)^2\right) \geq 0$.

Remark 2: To satisfy the above condition, the negative sign-in (27) must be changed to positive. The form (27) is determined as follows:

$$\dot{\tilde{\omega}}_{dq} = \gamma\sqrt{\left(\hat{x}_{kd}^2 + \hat{x}_{kq}^2\right)\left(\tilde{x}_{kd}^2 + \tilde{x}_{kq}^2\right) + \left(\hat{x}_{kd}\tilde{x}_{kd} + \hat{x}_{kq}\tilde{x}_{kq}\right)^2}. \tag{28}$$

Dependence (28) can be used to find the updated form of the classical estimation law (21). It provides an improvement to the stability range of the observer system.

Assumption 3. The expression (21) has the form of an open integrator. There is a lack of additional stabilizing function, interconnecting the observer system. To improve the stability range of the observer system, it is proposed to introduce additional input s_ω

$$\dot{\hat{\omega}}_{dq} = -\gamma\left(\hat{x}_{kq}\tilde{x}_{kd} - \hat{x}_{kd}\tilde{x}_{kq}\right) + s_{\omega}. \tag{29}$$

To stabilize the integrator (29), the stabilization function s_{ω} should be $s_{\omega} \equiv \dot{\hat{\omega}}_{dq}$. The updated estimation law has the following form:

$$\dot{\hat{\omega}}_{dq} = -\gamma\left(\hat{x}_{kq}\tilde{x}_{kd} - \hat{x}_{kd}\tilde{x}_{kq} + \gamma_1 k_f \sqrt{\left(\hat{x}_{kd}^2 + \hat{x}_{kq}^2\right)\left(\tilde{x}_{kd}^2 + \tilde{x}_{kq}^2\right) + \left(\hat{x}_{kd}\tilde{x}_{kd} + \hat{x}_{kq}\tilde{x}_{kq}\right)^2}\right), \tag{30}$$

where γ_1 is the additional gain, and $k_f = sign\left(\hat{\omega}_{dq}\right)$ is the sign of the estimated parameter.

Remark 3. Under the assumption that in (30) $\left(\hat{x}_{kd}^2 + \hat{x}_{kq}^2\right)\left(\tilde{x}_{kd}^2 + \tilde{x}_{kq}^2\right) \ll \left(\hat{x}_{kd}\tilde{x}_{kd} + \hat{x}_{kq}\tilde{x}_{kq}\right)^2$ the update estimation law can be simplified to the following form

$$\dot{\hat{\omega}}_{dq} = -\gamma\left(\hat{x}_{kq}\tilde{x}_{kd} - \hat{x}_{kd}\tilde{x}_{kq} + \gamma_1 k_f s_{\omega f}\right), \tag{31}$$

where $s_{\omega} = \hat{x}_{kd}\tilde{x}_{kd} + \hat{x}_{kq}\tilde{x}_{kq}$, and $s_{\omega f}$ is their filtered value by using a low-pass filter LPF (to avoid the algebraic loop).

In (31), there is the cross and scalar product $(\hat{\mathbf{x}}_k \cdot \tilde{\mathbf{x}}_k) = \hat{x}_{kd}\tilde{x}_{kd} + \hat{x}_{kq}\tilde{x}_{kq}$ of two vectors. It is worth noticing that for the perpendicular vectors, the scalar product is equal to zero; however, in other cases, it is different from zero and additionally stabilizes the estimation law.

2.2 Non-adaptive estimation of parameter ω_{dq}

In the previous section, the parameter ω_{dq} was reconstructed from the adaptive law. However, this value can be estimated non-adaptively. Under the assumption of the steady-state for $a_k \approx 1$, $\tilde{\omega}_{dq} \approx 0$ and $v_{d,\,q} = 0$, from the model of estimation error, the following approximations can be achieved:

$$\tilde{x}_{kd} \approx \hat{\omega}_{dq}\hat{x}_{kq}, \tag{32}$$

$$\tilde{x}_{kq} \approx -\hat{\omega}_{dq}\hat{x}_{kd}, \tag{33}$$

for whose the following relationships are satisfied

$$\tilde{x}_{kd}^2 + \tilde{x}_{kq}^2 = \hat{\omega}_{dq}^2\left(\hat{x}_{kd}^2 + \hat{x}_{kq}^2\right), \tag{34}$$

$$\hat{\omega}_{dq} = \frac{\tilde{x}_{kd}\hat{x}_{kq} - \tilde{x}_{kq}\hat{x}_{kd}}{\hat{x}_{kd}^2 + \hat{x}_{kq}^2}, \tag{35}$$

where $\hat{x}_{kd}^2 + \hat{x}_{kq}^2 \neq 0$.

Substituting (34)–(35) to (24), the following quadratic function is obtained:

$$f\left(\hat{\omega}_{dq}\right) = -\left(\hat{x}_{kd}^2 + \hat{x}_{kq}^2\right)\left(\hat{x}_{kd}^2 + \hat{x}_{kq}^2\right)\hat{\omega}_{dq}^2 + \hat{\omega}_{dq}\left(\hat{x}_{kd}^2 + \hat{x}_{kq}^2\right)\left(\tilde{x}_{kd}\hat{x}_{kq} - \tilde{x}_{kq}\hat{x}_{kd}\right) + \left(\hat{x}_{kd}\tilde{x}_{kd} + \hat{x}_{kq}\tilde{x}_{kq}\right)^2. \tag{36}$$

One of the roots of the function $f\left(\hat{\omega}_{dq}\right)$ can be calculated as follows:

$$\hat{\omega}_{dq} = \frac{\tilde{x}_{kd}\hat{x}_{kq} - \tilde{x}_{kq}\hat{x}_{kd} + k_f\sqrt{\left(\left(\tilde{x}_{kd}\hat{x}_{kq} - \tilde{x}_{kq}\hat{x}_{kd}\right)^2 + 4\gamma_1\left(\hat{x}_{kd}\tilde{x}_{kd} + \hat{x}_{kq}\tilde{x}_{kq}\right)^2\right)}}{2\left(\hat{x}_{kd}^2 + \hat{x}_{kq}^2\right)}, \tag{37}$$

where γ_1 is the additional tuning gain and $k_f = sign\left(\hat{\omega}_{dq}\right)$.

2.3 Practical stability of the observer system

The practical stability of the observer system was proposed in [17, 18]. Based on the theorem of practical stability and considering that the system belongs to domain $\boldsymbol{D}$ defined in (7), the observer structure will be practical stable in the Lyapunov function derivative is

$$\dot{V} = \delta_1|\tilde{x}_{kd}| + \delta_2|\tilde{x}_{kq}| + \delta_c|\tilde{\omega}_{dq}| \leq -\mu V + \kappa, \tag{38}$$

where $(\delta_1, \delta_2, \delta_c) > 0$ and $\tilde{x}_{kd} \leq \varepsilon_1$, $\tilde{x}_{kd} \leq \varepsilon_2$, $\tilde{\omega}_r \leq \varepsilon_3$, $\varepsilon_{1,2,3} \ll 1$ are sufficient small real numbers $\varepsilon_{1,2,3} > 0$ and where

$$\gamma_1 = \max\left\{\frac{\left(\hat{x}_{kq}\tilde{x}_{kd} - \hat{x}_{kd}\tilde{x}_{kq}\right)}{\sqrt{\left(\hat{x}_{kd}^2 + \hat{x}_{kq}^2\right)\left(\tilde{x}_{kd}^2 + \tilde{x}_{kq}^2\right) + \left(\hat{x}_{kd}\tilde{x}_{kd} + \hat{x}_{kq}\tilde{x}_{kq}\right)^2}} + \delta_c\right\}, \tag{39}$$

and

$$\mu = \min\left(\delta_1 - \frac{1}{2\xi_1^2}, \delta_2 - \frac{1}{2\xi_2^2}, \sqrt{2}\delta_c\right), \kappa = 0.5\left(\xi_1^2\eta_1^2 + \xi_2^2\eta_2^2\right) \ \forall \xi_i \in (0, 1), i = 1{,}2 \tag{40}$$

Hence, (38) implies the convergence of estimated vector values to their real, in finite time, noted as t_1. The reconstructed parameter $\hat{\omega}_{dq}$ converges exponentially to real ω_{dq} in finite time $t > t_2 > t_1$. This condition is satisfied for ideal and constant parameters of the system (3)–(4). According to [17, 18], the tracking errors converge to the ball of radius κ/μ. This radius can be decreased by the properly choosing tuning gains of the observer system (8)–(9).

2.4 Conclusion

Presented in Section 2 is the design of the observer structure generalized to the class of system (3)–(4) in the space vector form. The form of the system (3)–(4) was in α-β stationary reference frame. It has been appropriately transformed by using Clark's transformation to the rotational reference frame d-q. In system (5)–(6), there exists the parameter, which is the angular speed of the reference frame in d-q. The system (5)–(6) has been properly written with a separate parameter and has a similar form to an AC electrical machines models presented in the next sections. Therefore, the proposed procedure in Section 2 for designing the observer structure can be directly adapted to the sensorless control system of an AC electrical machine. The proposed solution is based on the classical adaptation law of estimation and non-adaptive. According to the literature, [1, 2, 6–12], the rotor speed whose value is estimated only from the classical law of adaptation and used to tune the observer structure (8)–(9) can lead to instability during the regenerating mode and low speed of the electrical machine. There exist positive poles of the observer structure for which it is unstable. The problem in the classical law of adaptation is the open form of the integrator (21) from which the value of rotor speed is estimated (in the case of an electrical machine). Therefore, in Section 2, the additional stabilization function is introduced also to the classical law of estimation. The proposed stabilization function is based on Lagrange's identity of the pair of vectors in the observer system. The form of additional stabilization law contains the scalar product and the length of the vectors. However, after the simplification shown in Remark 3, it can be assumed that the stabilization function is proportional to the scalar product of the chosen vectors that were presented in [6].

The proposed theoretical issues in Section 2 will be confirmed in the simulation and experimental results for the squirrel-cage induction machine and interior permanent magnet synchronous machine. Also, it can be extended to estimation of the state variables of the doubly fed induction generator (DFIG).

3. The speed observer structures of the squirrel-cage induction machine

The AFO speed observer structure of IM is proposed in this section. The rotor speed will be estimated by using two approaches: from the adaptive estimation law and non-adaptively.

Considering the mathematical model of the induction machine presented in [5, 6], for the pair of vectors (ψ_r, i_s) according to (8)–(9), the conventional AFO observer structure can be determined in the form

$$\frac{d\hat{i}_{s\alpha}}{d\tau} = a_1\hat{i}_{s\alpha} + a_2\hat{\psi}_{r\alpha} + a_3\hat{\omega}_r\hat{\psi}_{r\beta} + a_4u_{s\alpha} + v_\alpha, \tag{41}$$

$$\frac{d\hat{i}_{s\beta}}{d\tau} = a_1\hat{i}_{s\beta} + a_2\hat{\psi}_{r\beta} - a_3\hat{\omega}_r\hat{\psi}_{r\alpha} + a_4u_{s\beta} + v_\beta, \tag{42}$$

$$\frac{d\hat{\psi}_{r\alpha}}{d\tau} = a_5\hat{\psi}_{r\alpha} - \hat{\omega}_r\hat{\psi}_{r\beta} + a_6\hat{i}_{s\alpha} + v_{\psi\alpha}, \tag{43}$$

$$\frac{d\hat{\psi}_{r\beta}}{d\tau} = a_5\hat{\psi}_{r\beta} + \hat{\omega}_r\hat{\psi}_{r\alpha} + a_6\hat{i}_{s\beta} + v_{\psi\beta}, \tag{44}$$

where the estimated values are marked by "^".

It is assumed that the stator current vector $\hat{i}_{s\alpha,\beta}$, rotor flux vector $\hat{\psi}_{r\alpha,\beta}$ components, and rotor speed $\hat{\omega}_r$ are estimated in the observer structure (41)–(44), $v_{\alpha,\beta}$, and $v_{\psi\alpha,\beta}$ are stabilizing functions introduced to the structure. The values $i_{s\alpha,\beta}$ are available in measurement and $u_{s\alpha,\beta}$ are treated as the known variables (from the control system structure of the machine). The machine parameters are included in

$a_1 = -\frac{R_sL_r^2+R_rL_m^2}{L_rw_\sigma}$, $a_2 = \frac{R_rL_m}{L_rw_\sigma}$, $a_3 = \frac{L_m}{w_\sigma}$, $a_4 = \frac{L_r}{w_\sigma}$, $a_5 = -\frac{R_r}{L_r}$, $a_6 = \frac{R_rL_m}{L_r}$,

$$w_\sigma = L_r L_s - L_m^2. \tag{45}$$

The estimation errors for the observer system (41)–(44) are defined.

$$\tilde{\omega}_r = \hat{\omega}_r - \omega_r,,\tilde{\psi}_{r\alpha,\beta} = \hat{\psi}_{r\alpha,\beta} - \psi_{r\alpha,\beta}\tilde{i}_{s\alpha,\beta} = \hat{i}_{s\alpha,\beta} - i_{s\alpha,\beta} \tag{46}$$

where it is assumed that components $i_{s\alpha,\beta}$, $\psi_{r\alpha,\beta}$, ω_r are the real values.

The rotor speed value $\hat{\omega}_r$ will be reconstructed adaptively and non-adaptively by using the observer structure (41)–(44) and based on the measurements $i_{s\alpha,\beta}$, and $u_{s\alpha,\beta}$.

Using the design procedure presented in Section 2, the model of estimation errors is as follows:

$$\frac{d\tilde{i}_{s\alpha}}{d\tau} = a_1\tilde{i}_{s\alpha} + a_2\tilde{\psi}_{r\alpha} + a_3\left(\tilde{\omega}_r\hat{\psi}_{r\beta} + \hat{\omega}_r\tilde{\psi}_{r\beta} - \tilde{\omega}_r\tilde{\psi}_{r\beta}\right) + v_\alpha, \tag{47}$$

$$\frac{d\tilde{i}_{s\beta}}{d\tau} = a_1\tilde{i}_{s\beta} + a_2\tilde{\psi}_{r\beta} - a_3\left(\tilde{\omega}_r\hat{\psi}_{r\alpha} + \hat{\omega}_r\tilde{\psi}_{r\alpha} - \tilde{\omega}_r\tilde{\psi}_{r\alpha}\right) + v_\beta, \tag{48}$$

$$\frac{d\tilde{\psi}_{r\alpha}}{d\tau} = a_5\tilde{\psi}_{r\alpha} - \left(\tilde{\omega}_r\hat{\psi}_{r\beta} + \hat{\omega}_r\tilde{\psi}_{r\beta} - \tilde{\omega}_r\tilde{\psi}_{r\beta}\right) + a_6\tilde{i}_{s\alpha} + v_{\psi\alpha}, \tag{49}$$

$$\frac{d\tilde{\psi}_{r\beta}}{d\tau} = a_5\tilde{\psi}_{r\beta} + \left(\tilde{\omega}_r\hat{\psi}_{r\alpha} + \hat{\omega}_r\tilde{\psi}_{r\alpha} - \tilde{\omega}_r\tilde{\psi}_{r\alpha}\right) + a_6\tilde{i}_{s\beta} + v_{\psi\beta}. \tag{50}$$

The Lyapunov function defined for the estimation errors has the form

$$V = \frac{1}{2}\left(\tilde{i}_{s\alpha}^2 + \tilde{i}_{s\beta}^2 + \tilde{\psi}_{r\alpha}^2 + \tilde{\psi}_{r\beta}^2\right) + V_1 > 0, \tag{51}$$

where for the non-adaptive speed estimation $V_1 = 0$, and in (47)–(50), $\tilde{\omega}_r = 0$ under the assumption (32)–(33), for the case of adaptive law of estimation $V_1 = \frac{1}{\gamma}\tilde{\omega}_r^2$.

The derivative of the Lyapunov function will be negatively determined $\dot{V} < 0$ if the stabilizing functions are chosen.

$$v_\alpha = -c_\alpha\tilde{i}_{s\alpha},\; v_\beta = -c_\alpha\tilde{i}_{s\beta} \tag{52}$$

$$v_{\psi\alpha} = -c_{\psi 1}\tilde{i}_{s\alpha} + c_\psi\hat{\omega}_r\tilde{i}_{s\beta},\; v_{\psi\beta} = -c_{\psi 1}\tilde{i}_{s\beta} - c_\psi\hat{\omega}_r\tilde{i}_{s\alpha} \tag{53}$$

where it is assumed $(c_\alpha = f(a_1) > 0)$ as well as $c_\psi = f(a_3) > 0$, whereas $c_{\psi 1} \geq 0$, $a_5 < 0$.

The rotor speed value can be estimated directly from the adaptive estimation law (20) presented in Section 2.1, considering the pair of vectors $(\hat{\mathrm{x}}_k, \tilde{\mathrm{x}}_k) \equiv (\hat{\psi}_r, \tilde{\mathrm{i}}_s)$

$$\dot{\hat{\omega}}_r = -\gamma\left(\tilde{i}_{s\alpha}\hat{\psi}_{r\beta} - \tilde{i}_{s\beta}\hat{\psi}_{r\alpha} + \gamma_1 k_f \hat{s}_\omega\right), \tag{54}$$

where the robust term $\hat{s}_\omega$ is given from

$$\hat{s}_\omega = \sqrt{\left(\hat{\psi}_{r\alpha}^2 + \hat{\psi}_{r\beta}^2\right)\left(\tilde{i}_{s\alpha}^2 + \tilde{i}_{s\beta}^2\right) + \left(\hat{\psi}_{r\alpha}\tilde{i}_{s\alpha} + \hat{\psi}_{r\beta}\tilde{i}_{s\beta}\right)^2}, \tag{55}$$

and $\gamma_1 > 0$ is the additional gain, $k_f = sign(\hat{\omega}_r)$.

Considering the non-adaptive scheme for rotor speed estimation presented Section 2.2. For the pair of vectors $(\hat{x}_k, \tilde{x}_k) \equiv (\hat{\psi}_r, \tilde{i}_s)$, the rotor speed value can be estimated from

$$\hat{\omega}_r = \frac{\tilde{i}_{s\alpha}\hat{\psi}_{r\beta} - \tilde{i}_{s\beta}\hat{\psi}_{r\alpha} + k_f\hat{s}_\omega}{2\left(\hat{\psi}_{r\alpha}^2 + \hat{\psi}_{r\beta}^2\right)}, \tag{56}$$

where

$$\hat{s}_\omega = \sqrt{\left(\tilde{i}_{s\alpha}\hat{\psi}_{r\beta} - \tilde{i}_{s\beta}\hat{\psi}_{r\alpha}\right)^2 + 4\gamma_1\left(\hat{\psi}_{r\alpha}\tilde{i}_{s\alpha} + \hat{\psi}_{r\beta}\tilde{i}_{s\beta}\right)^2}. \tag{57}$$

The proposed AFO speed observer was tested on the 5.5 kW induction machine, which was clutched to DC motor. The sensorless control system structure was based on feedback control with the multi-scalar variables shown in [5, 6]. The control system contains four PI controllers of the rotor speed, electromagnetic torque T_e, square of rotor flux $\psi_r^2 = \psi_{r\alpha}^2 + \psi_{r\beta}^2$, and the variables $x_{22} = \hat{\psi}_{r\alpha}\hat{i}_{s\alpha} + \hat{\psi}_{r\beta}\hat{i}_{s\beta}$.

The control system was implemented in the interface with a DSP Sharc ADSP21363 floating-point signal processor with Altera Cyclone 2 FPGA. The interrupt time was 6.6 kHz, and the transistor switching frequency was 3.3 kHz. The rotor speed and position were measured by the incremental encoder (11-bits)—only to the accuracy verification of observer structure. The stator current was measured by the current transducers LA 25-NP—in the phases "a" and "b" and transformed to the (αβ) reference frame by using the Park transformation. The nominal parameters of the IM are presented in **Table 1**.

In **Figures 1–3**, the following variables are presented:

$\hat{\omega}_r$ - estimated rotor speed, ω_{rM} - measured rotor speed, $\tilde{\omega}_r$ - rotor speed error,$\hat{s}_\omega$ - additional variables, $\hat{\psi}_r^2 = \hat{\psi}_{r\alpha}^2 + \hat{\psi}_{r\beta}^2$, $\hat{T}_e = \hat{\psi}_{r\alpha}\hat{i}_{s\beta} - \hat{\psi}_{r\beta}\hat{i}_{s\alpha}$.

In **Figure 1**, the IM is starting up from 0.1 to 1.0 p.u. The waveforms of the estimated value of rotor speed, measured rotor speed, square of rotor flux

Parameter	Value	Unit
Nominal power	5.5	kW
Nominal speed	1430	rpm
Nominal voltage (Y)	400	V
Nominal current (Y)	11	A
Nominal frequency	50	Hz
Stator resistance R_{sN}	2.92/0.035	Ω/p.u.
Stator resistance R_{rN}	3.36/0.032	Ω/p.u.
Magnetizing inductance L_{mN}	0.422/1.95	H/p.u.
Stator and rotor inductance L_s, L_r	0.439/2.04	H/p.u.
$U_b = U_n$	0.82	p.u.
$I_b = I_n\sqrt{3}$	0.89	p.u.

Table 1.
Parameters of the IM and references unit.

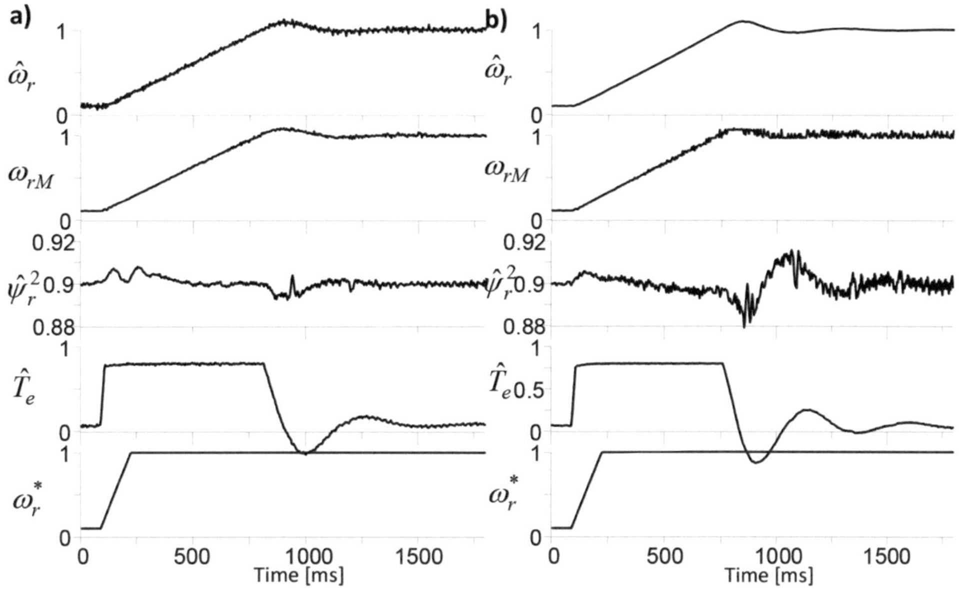

Figure 1.
The IM is starting up to 1.0 p.u., non-loaded and the rotor speed value is estimated from a) non-adaptive law (56), b) adaptively (54) – Experimental results.

components, electromagnetic torque, and reference rotor speed are presented. The reference value of electromagnetic torque is limited to 0.75 p.u. The reference value of the square of the rotor flux vector components is set to 0.9 p.u. The estimated rotor speed error during the dynamic states is about 0.015 p.u., and for the steady state is smaller than 0.01 p.u. In **Figure 1a**, the rotor speed is estimated non-adaptively. In **Figure 1b**, the rotor speed value is estimated from adaptive law.

In **Figure 2**, the rotor speed reversed from a nominal speed 1.0 p.u. to −1.0 p.u. The IM during this test is loaded at about T_L = 0.08 p.u. The square of rotor flux was set to 0.9 p.u. The value $\hat{s}_\omega$is determined from (57). The value of the rotor speed error for the case presented in **Figure 2a** is smaller than 0.02 in the dynamic states. For the case presented in **Figure 2b**, the value of rotor speed error is almost the same and smaller than 0.02 p.u.

In **Figure 3**, the motoring and regenerating modes of the IM are presented. In the AFO speed observer in which the rotor speed is estimated from the classical law of adaptation, for γ_1 = 0 in (54) (for this case, the stabilizing function is omitted), the observer structure is unstable in the regenerating machine mode, what was signaled in [6]. For the adaptive case (**Figure 3b**), if $\gamma_1 \neq 0$ and the value $\hat{s}_\omega$ is estimated from (55). The observer structure is stable during the load torque value change from 0.7 to −0.7 p.u. The rotor speed error is smaller than 0.015 in the dynamic states. The value of estimated electromagnetic torque for the regenerating case is about −0.65 p.u. It means that the electromagnetic torque value is estimated with a small value of the error of about 0.05 p.u. in the stationary state, but the observer structure is stable.

For the non-adaptive case presented in **Figure 3a**, the estimated rotor speed value has more oscillations than in **Figure 3a**. It is because the rotor speed is not filtered as in

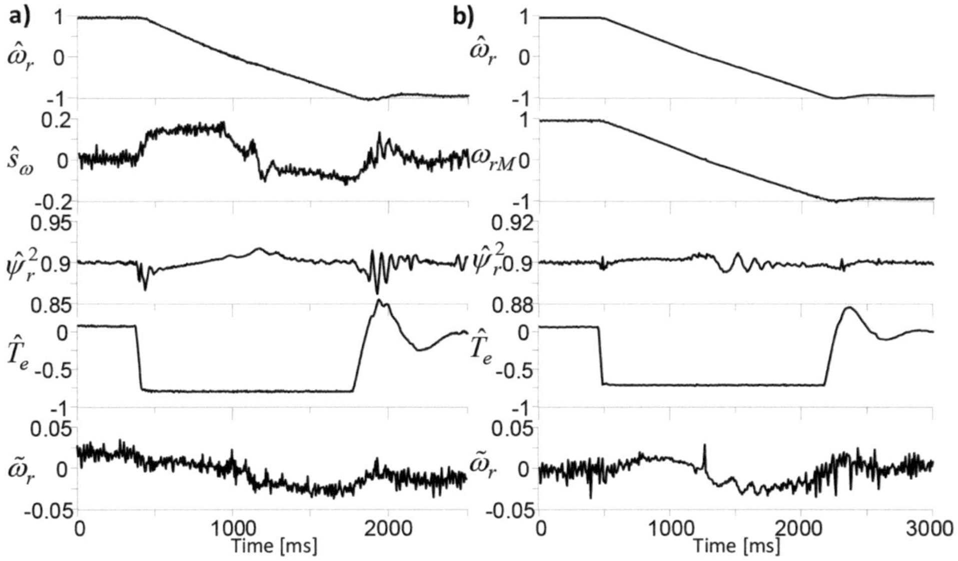

Figure 2.
The IM is reversing from 1.0 to −1.0 p.u., non-loaded and the rotor speed value is estimated from a) non-adaptive law (56), b) adaptively (54) – Experimental results.

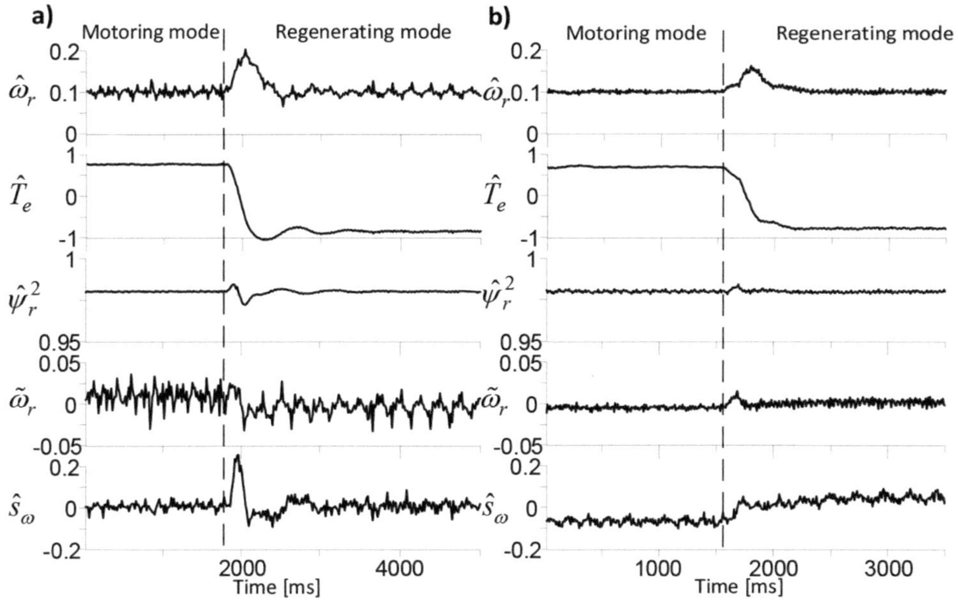

Figure 3.
Motoring and regenerating mode of IM for the rotor speed estimated a) from non-adaptive law (56), b) adaptively (54) – Experimental results.

the case of adaptive estimation law. However, the estimated value of electromagnetic torque is −0.7 p.u. (the same as the load torque). Hence, the electromagnetic torque is estimated more accurately than in the case of the adaptive law of rotor speed estimation.

3.1 Extended speed observer of the squirrel-cage induction machine

In [19], the speed observer was proposed, which is based on the extended model of the IM. In the model of the observer structure, an auxiliary variable marked in [5] as "Z" was introduced and defined as follows:

$$\hat{Z}_\alpha = \hat{\omega}_r \hat{\psi}_{r\alpha}, \tag{58}$$

$$\hat{Z}_\beta = \hat{\omega}_r \hat{\psi}_{r\beta}. \tag{59}$$

Based on the introduced auxiliary variables, the observer model can be determined

$$\frac{d\hat{i}_{s\alpha}}{d\tau} = a_1\hat{i}_{s\alpha} + a_2\hat{\psi}_{r\alpha} + a_3\hat{Z}_\beta + a_4 u_{s\alpha} + k_1(\hat{i}_{s\alpha} - i_{s\alpha}), \tag{60}$$

$$\frac{d\hat{i}_{s\beta}}{d\tau} = a_1\hat{i}_{s\beta} + a_2\hat{\psi}_{r\beta} - a_3\hat{Z}_\alpha + a_4 u_{s\beta} + k_1(\hat{i}_{s\beta} - i_{s\beta}), \tag{61}$$

$$\frac{d\hat{\psi}_{r\alpha}}{d\tau} = a_5\hat{\psi}_{r\alpha} - \hat{Z}_\beta + a_6\hat{i}_{s\alpha} + k_2(\hat{Z}_\beta - Z_\beta), \tag{62}$$

$$\frac{d\hat{\psi}_{r\beta}}{d\tau} = a_5\hat{\psi}_{r\beta} + \hat{Z}_\alpha + a_6\hat{i}_{s\beta} - k_2(\hat{Z}_\alpha - Z_\alpha), \tag{63}$$

$$\frac{d\hat{Z}_\alpha}{d\tau} = -\hat{\omega}_r(\hat{Z}_\beta - a_6 i_{s\alpha}) - a_5\hat{Z}_\alpha + k_3(\hat{i}_{s\alpha} - i_{s\alpha}), \tag{64}$$

$$\frac{d\hat{Z}_\beta}{d\tau} = \hat{\omega}_r(\hat{Z}_\alpha + a_6 i_{s\beta}) - a_5\hat{Z}_\beta + k_3(\hat{i}_{s\beta} - i_{s\beta}), \tag{65}$$

where the derivative of estimated rotor speed can be approximated $\frac{d\hat{\omega}_r}{d\tau} \approx \frac{\Delta\hat{\omega}_r}{\Delta T} \approx 0$ in the small interval time ΔT, and coefficients $a_1 \ldots a_6$ are defined in (45).

The rotor speed can be determined non-adaptively from the dependence [19]:

$$\hat{\omega}_r = \frac{\hat{Z}_\alpha\hat{\psi}_{r\alpha} + \hat{Z}_\beta\hat{\psi}_{r\beta}}{\hat{\psi}_{r\alpha}^2 + \hat{\psi}_{r\beta}^2}. \tag{66}$$

The experimental results in this section are limited only to the regenerating mode of the IM, in which the observer structure can be unstable. The reference rotor speed is set to 0.1 p.u.

In the first case presented in **Figure 4a**, (in which the rotor speed is estimated from (66)) after 0.5 s machine is loaded $T_L = -0.6$ p.u. For the motoring mode ($\hat{\omega}_r > 0$, $\hat{T}_e \geq 0$), the speed observer (60)–(65) is stable. After 1.5 s, when the load torque value is decreased up to -0.2 p.u., the observer system estimates the state variables incorrectly, the error of estimated rotor speed increases up to 0.05 p.u., and after 1.8 s, the electromagnetic torque value achieves its limitation (0.75 p.u.). After 1.9 s, the rotor speed error is higher than 0.05 p.u., and the IM is braking. The observer structure achieves unstable points of operation in which all the estimates do not converge to their real values.

The rotor speed value is estimated from (66), which is suitable only for the motoring mode of the machine. This is the same case as for the AFO speed observer structure. The speed estimation law is based on the algebraic eq. (66), which does not

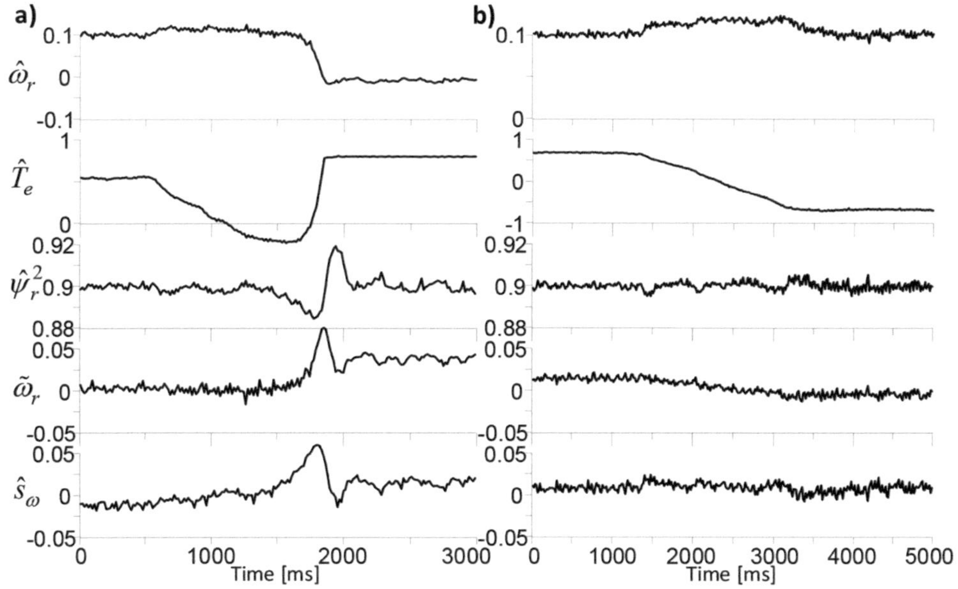

Figure 4.
Regenerating mode of IM for the rotor speed estimated: a) the rotor speed value is estimated from (66), b) the rotor speed is estimated by using the proposed robust law (additional stabilization function) – Experimental results.

guarantee the stability of the observer structure during the regenerating mode of the machine. Some poles of the observer move to an unstable zone (are zero or positive). The reason for this is the form of dependence (66) in which the additional stabilization function does not exist. The stabilization function, proposed in Section 2.2, which is based on Lagrange's identity, cannot be directly used, because the vectors: $\hat{\psi}_r$ and $\hat{Z}$have the same position and different amplitude only. This is the result of the definition (58)–(59). In this case, it is better to use from (58)–(59), and after few simple transformations, one can be obtain

$$\hat{Z}_\alpha \hat{\psi}_{r\beta} - \hat{Z}_\beta \hat{\psi}_{r\alpha} = \hat{\omega}_r \hat{\psi}_{r\alpha} \hat{\psi}_{r\beta} - \hat{\omega}_r \hat{\psi}_{r\alpha} \hat{\psi}_{r\beta} = 0. \tag{67}$$

This is satisfied for the ideal case in which all estimation errors are equal to zero. In the other case, taking the left side of (67) as

$$\hat{s}_\omega = \hat{Z}_\alpha \hat{\psi}_{r\beta} - \hat{Z}_\beta \hat{\psi}_{r\alpha}, \tag{68}$$

and using (66) the update form of non-adaptive speed estimation with the stabilization function (68) can be determined

$$\hat{\omega}_r = \frac{\hat{Z}_\alpha \hat{\psi}_{r\alpha} + \hat{Z}_\beta \hat{\psi}_{r\beta} + k_f \gamma_1 \hat{s}_\omega}{\hat{\psi}_{r\alpha}^2 + \hat{\psi}_{r\beta}^2}, \tag{69}$$

where $k_f = sign(\hat{\omega}_r)$, $\gamma_1 \geq 0$.

The experimental results of the proposed non-adaptive speed estimation with the stabilization function (68) are presented in **Figure 4b**. The reference of rotor speed is set to 0.1 p.u. After 1.5 s, the load torque slowly changes from (the motoring mode)

0.7 to −0.7 p.u. (the regenerating mode). The speed observer correctly estimates the electromagnetic torque value with the rotor speed error smaller than 0.015 p.u. during the dynamic states. The square of rotor flux vector components is stabilized on the almost constant value equal to 0.9 p.u. The proposed stabilized function (68) improves the speed observer properties making the speed observer structure more robust in the regenerating mode, which was confirmed in **Figure 4b**.

4. Speed and position observer of interior permanent magnet machine

This section is concerned with the observer system of interior permanent magnet machines IPMSM and their problems during a sensorless application under disturbances.

4.1 Mathematical models of the IPMSM machines

The mathematical model of IPMSM is often determined in the rotating reference frame (d-q), which is connected to the position of the rotor. The model in (d-q) was presented in [12–14]. However, sometimes the mathematical model is better considered in the stationary (α-β) reference frame connected to the stator. The model of IPMSM in (α-β) has the following form [12]:

$$\frac{di_{s\alpha}}{d\tau} = \frac{\omega_r}{L_d}\lambda_\beta + (-R_s i_{s\alpha} + u_{s\alpha})L_1 + \left(-R_s i_{s\beta} + u_{s\beta}\right)L_3, \tag{70}$$

$$\frac{di_{s\beta}}{d\tau} = -\frac{\omega_r}{L_d}\lambda_\alpha + (-R_s i_{s\alpha} + u_{s\alpha})L_3 + \left(-R_s i_{s\beta} + u_{s\beta}\right)L_4, \tag{71}$$

$$\frac{d\omega_r}{d\tau} = \frac{1}{J}\left(\psi_{f\alpha} i_{s\beta} - \psi_{f\beta} i_{s\alpha} + \left(L_d - L_q\right) i_{s\alpha} i_{s\beta} - T_L\right), \tag{72}$$

$$\frac{d\theta_r}{d\tau} = \omega_r, \tag{73}$$

where

$$\lambda_\alpha = L_d L_q^{-1}\psi_{f\alpha} - \left(1 - L_d L_q^{-1}\right)(L_0 i_{\alpha 2} + L_2 i_{s\alpha}), \tag{74}$$

$$\lambda_\beta = L_d L_q^{-1}\psi_{f\beta} + \left(1 - L_d L_q^{-1}\right)\left(L_0 i_{\beta 2} - L_2 i_{s\beta}\right), \tag{75}$$

$$L_0 = 0.5\left(L_d + L_q\right),, L_1 = L_d^{-1}\cos^2\theta_r + L_q^{-1}\sin^2\theta_r \tag{76}$$

$$L_2 = 0.5\left(L_d - L_q\right),, L_3 = 0.5\left(\frac{1}{L_d} - \frac{1}{L_q}\right)\sin(2\theta_r) \tag{77}$$

$$L_4 = L_d^{-1}\sin^2\theta_r + L_q^{-1}\cos^2\theta_r, \tag{78}$$

$$i_{\alpha 2} = i_{s\alpha}\cos 2\theta_r + i_{s\beta}\sin 2\theta_r \tag{79}$$

$$i_{\beta 2} = -i_{s\alpha}\sin 2\theta_r + i_{s\beta}\cos 2\theta_r, \tag{80}$$

$$\psi_{f\alpha} = \psi_f\cos\theta_r,, \psi_{f\beta} = \psi_f\sin\theta_r \tag{81}$$

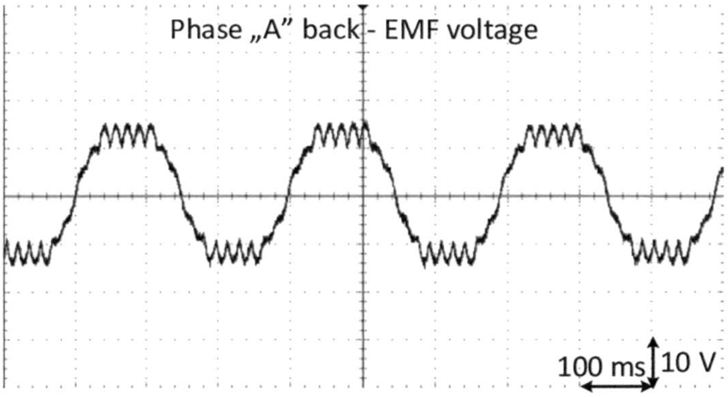

Figure 5.
The phase "A" back-EMF voltage.

where R_s is the stator resistance, L_d, L_q are the winding inductances, $i_{s\alpha,\beta}$ are the stator currents, $u_{s\alpha,\beta}$ are the stator voltages, ψ_f is the permanent magnet flux linkage, ω_r is the rotor speed, θ_r is the rotor position, J is the rotor inertia, T_L is the load torque.

The design of the IPMSM machine has a significant impact on the properties of the whole drive system. In IPMSM without the skews in the slots, there occur the rotor slot's harmonics, [20]. The slot's harmonics cause the non-sinusoidal distribution of the electromagnetic force (EMF) generated in the machine, [21]. These have a negative influence on the quality of the control system of the machine, and in particular, on the speed and position observer. In the literature [12–14], these negative effects are named by the disturbances in the IPMSM, which have bounded values. These disturbances have a significant impact on the machine rotor speed smaller than 15% of the nominal value and the idling mode of the IPMSM. For the low-speed range, the stator voltage has a small value, and similarly is the stator currents; because of this, the back-EMF value is significant. The example waveform of back-EMF voltage in 3.5 kW IPMSM machine for 10% of nominal rotor speed, registered by using the digital oscilloscope is presented in **Figure 5**.

In **Figure 5**, there are visible 18 slot's harmonics in the waveform. The IPMSM nominal parameters are shown in **Table 2**.

Parameter	Value	Unit
Nominal power	3.5	kW
Nominal speed	1500	rpm
Nominal voltage (Y)	285	V
Nominal current (Y)	7.5	A
Stator resistance R_s	0.023	p.u.
Inductance L_{dN}	0.28	p.u.
Inductance L_{qN}	0.82	p.u.
Rotor flux linkage	0.89	p.u.

Table 2.
IPMSM nominal parameters and reference units.

In the next section, the speed and position observer is proposed for the IPMSM in which the disturbances have occurred and the back-EMF voltage has an almost trapezoidal distribution (**Figure 5**).

4.2 Adaptive speed and position observer of IPMSM

As was mentioned in 4.1, the IPMSM has disturbances in the form of trapezoidal back-EMF voltage with slot's harmonics. The classical structures of the observer in (d-q) are not stable if the rotor speed is smaller than 15–20% of the nominal speed. One of the solutions to overcome this problem is to implement a dedicated algorithm in which additional high or low-frequency signals are injected into the stator voltage or current [13, 14]. However, the sensorless control system is then more complicated than classical FOC, and the observer structure contains a few low-passes or bandwidth filters [22]. The procedure of selecting the settings of the observer and the PI controllers in the control system is difficult. Therefore in this section, a new form of the speed and position observer is proposed, which is based on the mathematical model in the stationary reference frame presented in Section 4.1. Considering the procedure of design of the observer stabilization function from Section 2, the AFO speed observer of IPMSM can be determined

$$\frac{d\hat{i}_{s\alpha}}{d\tau} = \frac{\hat{\omega}_r}{L_d}\hat{\lambda}_\beta + \left(-R_s\hat{i}_{s\alpha} + u_{s\alpha}\right)L_1 + \left(-R_s\hat{i}_{s\beta} + u_{s\beta}\right)L_3 + v_\alpha, \tag{82}$$

$$\frac{d\hat{i}_{s\beta}}{d\tau} = -\frac{\hat{\omega}_r}{L_d}\hat{\lambda}_\alpha + \left(-R_s\hat{i}_{s\alpha} + u_{s\alpha}\right)L_3 + \left(-R_s\hat{i}_{s\beta} + u_{s\beta}\right)L_4 + v_\beta, \tag{83}$$

$$\frac{d\hat{\theta}_r}{d\tau} = \hat{\omega}_r + v_\theta, \tag{84}$$

where "^" denotes estimated values; $v_{\alpha,\beta,}$ and v_θ are stabilizing functions introduced to (82)–(83).

The values of the rotor flux vector components can be obtained by using (74)–(75) as follows:

$$\hat{\lambda}_\alpha = L_d L_q^{-1}\hat{\psi}_{f\alpha} - \left(1 - L_d L_q^{-1}\right)\left(L_0\hat{i}_{\alpha 2} + L_2\hat{i}_{s\alpha}\right), \tag{85}$$

$$\hat{\lambda}_\beta = L_d L_q^{-1}\hat{\psi}_{f\beta} + \left(1 - L_d L_q^{-1}\right)\left(L_0\hat{i}_{\beta 2} - L_2\hat{i}_{s\beta}\right), \tag{86}$$

where

$$L_0 = 0.5\left(L_d + L_q\right),, L_1 = L_d^{-1}\cos^2\hat{\theta}_r + L_q^{-1}\sin^2\hat{\theta}_r \tag{87}$$

$$L_2 = 0.5\left(L_d - L_q\right),, L_3 = 0.5\left(\frac{1}{L_d} - \frac{1}{L_q}\right)\sin\left(2\hat{\theta}_r\right) \tag{88}$$

$$L_4 = L_d^{-1}\sin^2\hat{\theta}_r + L_q^{-1}\cos^2\hat{\theta}_r, \tag{89}$$

$$\hat{i}_{\alpha 2} = \hat{i}_{s\alpha}\cos 2\hat{\theta}_r + \hat{i}_{s\beta}\sin 2\hat{\theta}_r \tag{90}$$

$$\hat{i}_{\beta 2} = -\hat{i}_{s\alpha}\sin 2\hat{\theta}_r + \hat{i}_{s\beta}\cos 2\hat{\theta}_r, \tag{91}$$

$$\hat{\psi}_{f\alpha} = \psi_f \cos\hat{\theta}_r, . \hat{\psi}_{f\beta} = \psi_f \sin\hat{\theta}_r \tag{92}$$

To stabilize the observer structure (82)–(84), the appropriate form of the stabilization functions $v_{\alpha,\beta}$ and v_θ should be determined to satisfy the Lyapunov theorem. The Lyapunov candidate function has the form

$$V = 0.5\left(\left(\tilde{i}_{s\alpha}^2 + \tilde{i}_{s\beta}^2\right) + \tilde{\theta}_r^2 + \gamma^{-1}\tilde{\omega}_r^2\right), \tag{93}$$

where

$$\tilde{i}_{s\alpha,\beta} = \hat{i}_{s\alpha,\beta} - i_{s\alpha,\beta}, , . \tilde{\omega}_r = \hat{\omega}_r - \omega_r \tilde{\theta}_r = \hat{\theta}_r - \theta_r \tag{94}$$

The derivative of the Lyapunov function can be determined by using the estimation errors (94) and the proposed observer structure as

$$\dot{V} = -\left(c_\alpha \tilde{i}_{s\alpha}^2 + c_\alpha \tilde{i}_{s\beta}^2\right) + \tilde{\omega}_r\left(-\frac{1}{L_d}\hat{\lambda}_\beta \tilde{i}_{s\alpha} + \frac{1}{L_d}\hat{\lambda}_\alpha \tilde{i}_{s\beta} + \frac{1}{\gamma}\dot{\hat{\omega}}_r\right) + \tilde{\theta}_r(\tilde{\omega}_r + v_\theta) \leq 0. \tag{95}$$

The derivative of the Lyapunov function (95) is negative if the stabilizing functions are chosen

$$v_\alpha = -c_\alpha R_s L_1 \tilde{i}_{s\alpha} + c_\lambda L_d^{-1} \hat{\omega}_r \hat{\lambda}_\beta \tilde{i}_{s\alpha}, \tag{96}$$

$$v_\beta = -c_\alpha R_s L_4 \tilde{i}_{s\beta} - c_\lambda L_d^{-1} \hat{\omega}_r \hat{\lambda}_\alpha \tilde{i}_{s\beta}, \tag{97}$$

$$v_\theta = -c_\theta \tilde{\theta}_r, \tag{98}$$

where $(c_\alpha, c_\theta) > 0$ and $c_\lambda \leq \frac{R_s L_1 \hat{\lambda}_\beta - R_s L_4 \hat{\lambda}_\alpha}{L_d^{-1}|\hat{\omega}_r|\left(\hat{\lambda}_\alpha^2 + \hat{\lambda}_\beta^2\right)}$.

The rotor speed value can be estimated by using the classical adaptation law

$$\dot{\hat{\omega}}_r = \gamma L_d^{-1}\left(\hat{\lambda}_\beta \tilde{i}_{s\alpha} - \hat{\lambda}_\alpha \tilde{i}_{s\beta}\right), \tag{99}$$

where $\gamma > 0$.

However, under *Assumption 3* from Section 2.1 in order to improve the quality of reconstruction of the rotor speed value, it is better to introduce the additional stabilization function to the open-integrator (99)

$$\dot{\hat{\omega}}_r = \gamma L_d^{-1}\left(\hat{\lambda}_\beta \tilde{i}_{s\alpha} - \hat{\lambda}_\alpha \tilde{i}_{s\beta} + k_f \hat{s}_\omega\right), \tag{100}$$

where the value of the stabilizing function can be obtained by using the approach presented in Section 2.1:

$$\hat{s}_\omega = \sqrt{\left(\hat{\lambda}_\alpha^2 + \hat{\lambda}_\beta^2\right)\left(\tilde{i}_{s\alpha}^2 + \tilde{i}_{s\beta}^2\right) + \left(\hat{\lambda}_\alpha \tilde{i}_{s\alpha} + \hat{\lambda}_\beta \tilde{i}_{s\beta}\right)^2}. \tag{101}$$

For $\left(\hat{\lambda}_\alpha^2 + \hat{\lambda}_\beta^2\right)\left(\tilde{i}_{s\alpha}^2 + \tilde{i}_{s\beta}^2\right) \ll \left(\hat{\lambda}_\alpha \tilde{i}_{s\alpha} + \hat{\lambda}_\beta \tilde{i}_{s\beta}\right)^2$, the value of (101) can be determined from the simplified form

$$\hat{s}_\omega = \hat{\lambda}_\alpha \tilde{i}_{s\alpha} + \hat{\lambda}_\beta \tilde{i}_{s\beta}. \tag{102}$$

The rotor position can be obtained directly from (84), and the stabilizing function v_θ from (98).

In (98) there is the rotor position error, which is defined $\tilde{\theta}_r = \hat{\theta}_r - \theta_r$, where θ_r means the real (measured) value of rotor speed. However, in the speed observer structure, the rotor speed is not measured but only estimated. Therefore, it is proposed to replace the deviation $\tilde{\theta}_r$ by $\tilde{\theta}_\lambda$ and (98) is rewritten as

$$v_\theta = -c_\theta \tilde{\theta}_\lambda, \tag{103}$$

where $\tilde{\theta}_\lambda$ can be defined as the angle between the rotor flux vector components $\lambda_{\alpha,\beta}$ and their estimated values $\hat{\lambda}_{\alpha,\beta}$. Values of deviation $\tilde{\theta}_\lambda$ can be determined as

$$\tilde{\theta}_\lambda = \tan^{-1}(\varphi), \tag{104}$$

where $\varphi = \left(\lambda_\alpha \hat{\lambda}_\beta - \lambda_\beta \hat{\lambda}_\alpha\right)\left(\lambda_\alpha \hat{\lambda}_\alpha + \lambda_\beta \hat{\lambda}_\beta\right)^{-1}$. The rotor flux vector components, $\lambda_{\alpha,\beta}$, can be determined from (74)–(75) in which it is assumed $\theta_r \approx \hat{\theta}_r$ and the measured values of $i_{s\alpha,\beta}$ are used; also, $\left(\lambda_\alpha \hat{\lambda}_\alpha + \lambda_\beta \hat{\lambda}_\beta\right) \neq 0$.

Value of $\tilde{\theta}_\lambda$ should be projected using

$$\tilde{\theta}_\lambda = \begin{Bmatrix} \tilde{\theta}_\lambda - \pi/2, \text{ } if \text{ } \varphi > 0 \\ \tilde{\theta}_\lambda + \pi/2 \text{ } if \text{ } \varphi < 0 \end{Bmatrix}, \tag{105}$$

It gives the values $\tilde{\theta}_\lambda$ in a steady state close to zero, and it can be assumed that $\tilde{\theta}_\lambda \approx \tilde{\theta}_r$. The proposed stabilizing function improves the estimated value of the rotor position, particularly in the dynamic states of the IPMSM. The stabilizing function is necessary in the case of IPMSM with the described above disturbances.

Remark 4. The value of rotor flux vector components must be estimated from (74)–(75), however, by using the estimated rotor speed position $\hat{\theta}_r$ in (76)–(81).

4.3 Non-adaptive speed estimation of the IPMSM

The rotor speed value can be estimated by using the non-adaptive estimation scheme. Consider the non-adaptive scheme for the rotor speed estimation presented in Section 2.2 for the pair of vectors, $(\hat{\mathrm{x}}_k, \tilde{\mathrm{x}}_k) \equiv \left(\hat{\lambda}_r, \tilde{\mathrm{i}}_s\right)$ the rotor speed value can be estimated from

$$\hat{\omega}_r = \frac{\tilde{i}_{s\alpha}\hat{\lambda}_\beta - \tilde{i}_{s\beta}\hat{\lambda}_\alpha + k_f \hat{s}_\omega}{2\left(\hat{\lambda}_\alpha^2 + \hat{\lambda}_\beta^2\right)}, \tag{106}$$

where

$$\hat{s}_\omega = \sqrt{\left(\tilde{i}_{s\alpha}\hat{\lambda}_\beta - \tilde{i}_{s\beta}\hat{\lambda}_\alpha\right)^2 + 4\gamma_1\left(\hat{\lambda}_\alpha \tilde{i}_{s\alpha} + \hat{\lambda}_\beta \tilde{i}_{s\beta}\right)^2}, \tag{107}$$

and $k_f = sign(\hat{\omega}_r)$, $\gamma_1 \geq 0$.

4.4 Simulation and experimental results of the speed and position observer of IPMSM

In this section, the chosen waveforms from the simulation and the experiment setup are shown. The nominal parameters of the IPMSM are shown in **Table 2**. The experimental validations were carried out on 3.5 kW IPMSM. The stator of IPMSM has 18 slots, which are visible in the waveform of EMF from **Figure 5**. The machine is controlled by using the classical FOC control presented in [12, 13, 22]. There are three PI controllers for the rotor speed, i_{sq}, and i_{sd} stator vector components. Additionally, the MTPA algorithm [12] was applied.

In **Figure 6,** the waveform of the simulation results is shown. The estimated rotor speed $\hat{\omega}_r$, stator current vector components $\hat{i}_{sd,q}$ estimated rotor speed error $\tilde{\omega}_r$, and the estimated rotor position $\hat{\theta}_r$ are presented. In **Figure 6a**, the machine is starting up to 1.0 p.u. and after 600 ms loaded to about 0.6 p.u. The error of the rotor position is smaller than 0.05 p.u. during the dynamic states, the rotor speed error is smaller than 0.01 p.u. In **Figure 6b**, the machine is reversing to −1.0 p.u. The position error is smaller than 0.1 p.u. during the dynamic states, and the error of rotor speed is smaller than 0.05 p.u. In **Figure 6b**, the measured value of rotor position θ_{rM} is shown.

In **Figure 7a**, after 100 ms the machine is loaded T_L = 1.0 p.u. and after 600 ms the regenerating mode is applied and T_L = −1.0 p.u. The rotor reference speed is equal to 0.1 p.u. The estimated electromagnetic torque $\hat{T}_e$, rotor position error $\tilde{\theta}_r$, $\tilde{\theta}_\lambda$ defined in (104), s_ω and rotor speed error, and the estimated $\hat{i}_{sd}$ stator current component are presented. It is worth noticing that the waveforms of $\tilde{\theta}_r$ as well as $\hat{s}_\omega$ and $\tilde{\omega}_r$ are converged on each other.

The experimental waveforms are presented in **Figure 8**. The machine's reversal from 1.0 to −1.0 p.u. is shown in **Figure 8a**. The estimated rotor speed $\hat{\omega}_r$, stator current vector components $\hat{i}_{sd,q}$ estimated rotor speed error $\tilde{\omega}_r$, and the estimated

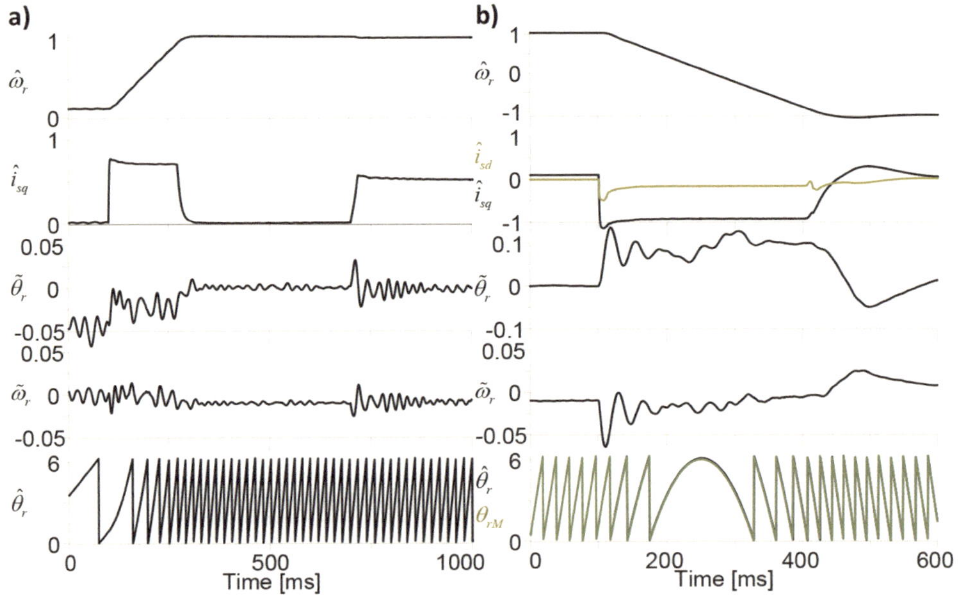

Figure 6.
a) Machine is starting up to 1.0 p.u and b) reversing to −1.0 p.u. – Simulation results.

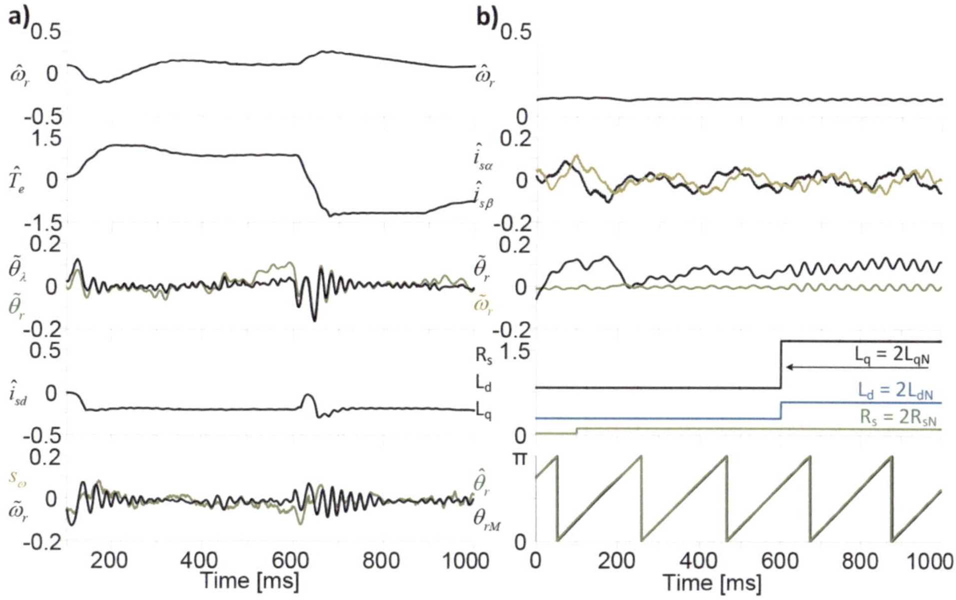

Figure 7.
a) the load torque T_L *is changed from 1.0 to – 1.0 p.u., b) parameters of the machine are changed in the sensorless control system (parameters uncertainties test) – Simulation.*

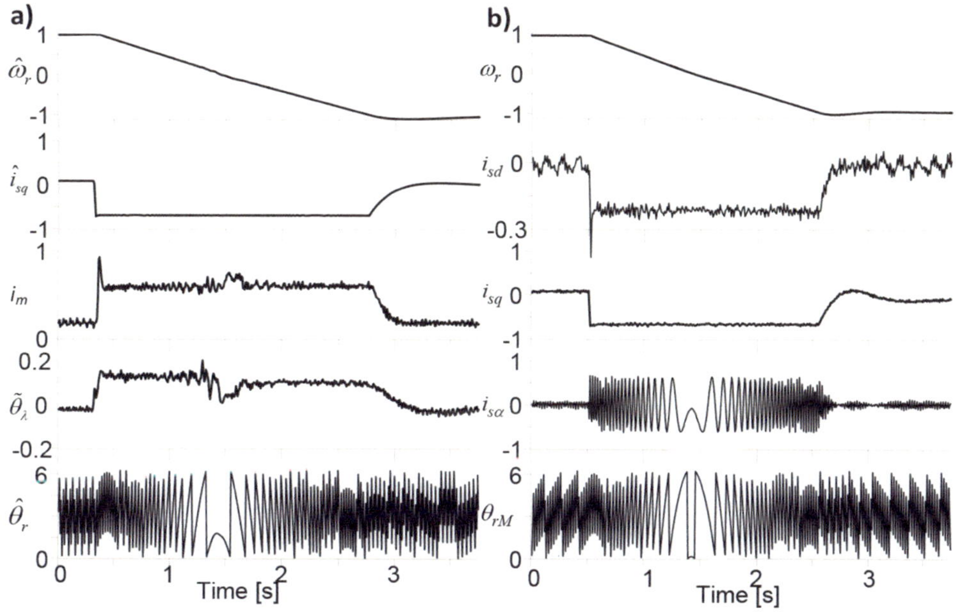

Figure 8.
Machine is reversing from 1.0 to -1.0 *p.u. for a) the sensorless control system with the proposed observer, b) the control system with measured rotor speed and position values – Experimental results.*

rotor position $\hat{\theta}_r$ and the stator current module i_m are presented. In **Figure 8b**, the same waveforms but for the measured value from the encoder of the rotor speed and rotor position are presented (for comparison). During the machine reverse, the i_{sd} value is about 0.25 p.u. It results from the MTPA algorithm [12].

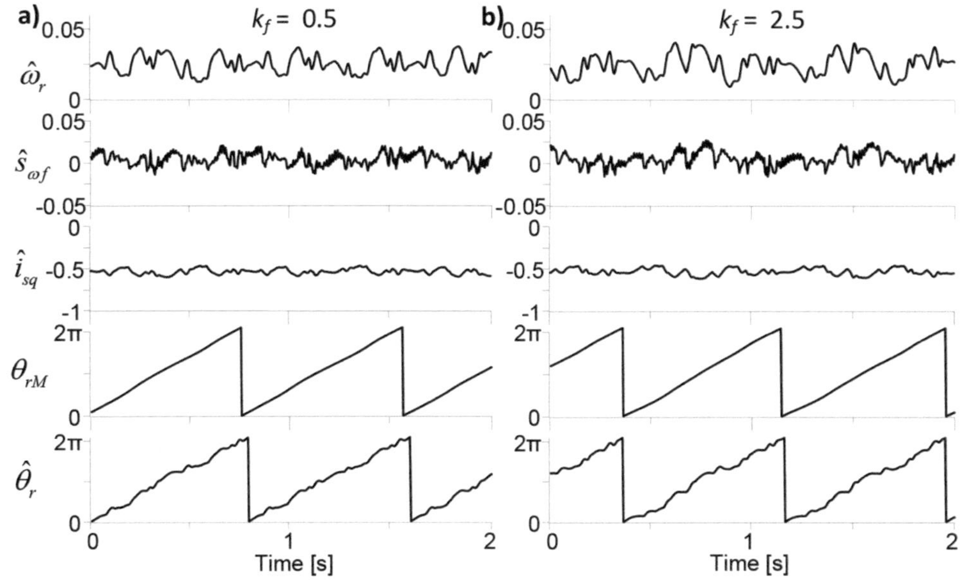

Figure 9.
Regenerating mode of IPMSM machine for the low reference speed 0.025 p.u., $T_L = -0.5$ *p.u. for the cases a) in (106)* $k_f = 0.5$*, b) in (106)* $k_f = 2.5$ *p.u. – Experimental results.*

In **Figure 9**, the regenerating mode of IPMSM is shown. The rotor speed was set to 0.025 p.u., and the machine was loaded at about −0.5 p.u. In **Figure 9a**, the stabilizing function is $k_f = 0.5$ in (106), and the value of the error of estimated rotor position is increased to 0.075 p.u., which is visible in **Figure 9a**. The value of the stator current component $\hat{i}_{sq}$was incorrectly estimated (due to rotor position error). In **Figure 9b**, the same case is shown, however for $k_f = 2.5$ p.u. The rotor position error is almost minimized to zero, and the $\hat{i}_{sq}$ value is about −0. 5 p.u. (the same as the referenced).

In this section, the presented simulation and experimental results confirmed that the introduced stabilization function into the speed adaptation scheme leads to the improvement of the properties of the observer system and robustness of the occurred disturbances. In this case, these are the non-sinusoidal EMF and slot's harmonics.

5. Speed and position observer of doubly fed induction generator

In this section, the speed and position observer of the doubly fed induction generator DFIG is considered. The rotor is connected to a voltage source converter, and the stator is directly connected to three phases AC-grid. The field-oriented control FOC is used to control the active and reactive stator powers presented in [23]. The rotor speed will be estimated by using a non-adaptive estimation scheme only.

5.1 Mathematical models of the DFIG

The mathematical model of DFIG can be determined in rotating or stationary reference frames (x-y). Considering the rotor and stator current vector components, the differential equations have the form [24]:

$$\frac{di_{sx}}{d\tau} = -\frac{L_s}{w_\sigma}(R_s i_{sx} - u_{sx}) + \frac{L_m}{w_\sigma}(\omega_r(L_m i_{sy} + L_r i_{ry}) + R_r i_{rx} - u_{rx}), \tag{108}$$

$$\frac{di_{sy}}{d\tau} = -\frac{L_r}{w_\sigma}(R_s i_{sy} - u_{sy}) - \frac{L_m}{w_\sigma}(\omega_r(L_m i_{sx} + L_R i_{rx}) - R_r i_{ry} + u_{ry}), \tag{109}$$

$$\frac{di_{rx}}{d\tau} = \frac{L_s}{w_\sigma}(-\omega_r(L_r i_{ry} + L_m i_{sy}) - R_r i_{rx} + u_{rx}) + \frac{L_m}{w_\sigma}(R_s i_{sx} - u_{sx}), \tag{110}$$

$$\frac{di_{ry}}{d\tau} = \frac{L_s}{w_\sigma}(\omega_r(L_r i_{rx} + L_m i_{sx}) - R_r i_{ry} + u_{ry}) + \frac{L_m}{w_\sigma}(R_s i_{sy} - u_{sy}), \tag{111}$$

$$\frac{d\omega_r}{d\tau} = \frac{L_m}{JL_r}(i_{rx} i_{sy} - i_{ry} i_{sx}) - \frac{1}{J}(T_L + f_r \omega_r). \tag{112}$$

where the (x-y) coordinate system is associated with any angular speed, and it is assumed that (108)–(109) are connected to the stationary stator windings so the angular speed of the (x,y) system is $\omega_a = 0, f_r$ is the friction.

It is assumed that all the DFIG parameters are known and constant. The components u_{rx}, u_{ry} are treated as the control vector variables, and u_{sx}, u_{sy}, i_{sx}, i_{sy} and i_{rx}, i_{ry} components are treated as measured and transformed to the adequate (x-y) reference frame.

To design the observer structure, it is proposed to introduce new auxiliary variables, which are defined

$$H_x = \omega_r(L_m i_{sx} + L_r i_{rx}), \tag{113}$$

$$H_y = \omega_r(L_m i_{sy} + L_r i_{ry}). \tag{114}$$

Considering (113)–(114) and (108)–(111), the observer structure can be determined

$$\frac{d\hat{i}_{rx}}{d\tau} = -\frac{L_s}{w_\sigma}(\hat{H}_y + R_r i_{rx} + u_{rx}) + \frac{L_m}{w_\sigma}(R_s i_{sx} - u_{sx}) + v_{rx}, \tag{115}$$

$$\frac{d\hat{i}_{ry}}{d\tau} = \frac{L_s}{w_\sigma}(\hat{H}_x - R_r i_{ry} + u_{ry}) + \frac{L_m}{w_\sigma}(R_s i_{sy} - u_{sy}) + v_{ry}, \tag{116}$$

$$\frac{d\hat{H}_x}{d\tau} = \hat{\omega}_r(-\hat{H}_y - R_r i_{rx} + u_{rx}) + v_{Hx}, \tag{117}$$

$$\frac{d\hat{H}_y}{d\tau} = \hat{\omega}_r(\hat{H}_x - R_r i_{ry} + u_{ry}) + v_{Hy}, \tag{118}$$

$$\frac{d\hat{\theta}_r}{d\tau} = \hat{\omega}_r + v_\theta, \tag{119}$$

where estimated state variables are marked by "^" and $\frac{d\hat{\omega}_r}{d\tau} \approx \frac{\Delta\hat{\omega}_r}{\Delta T}$, the observer contains the stabilization functions v_{rx}, v_{ry} and v_{Hx}, v_{Hy}, v_θ .

According to the design procedure of the observer presented in Section 2, in order to stabilize the observer structure (115)–(119), appropriate form of the stabilization functions v_{rx}, v_{ry} and v_{Hx}, v_{Hy}, v_θ should be determined to satisfy the Lyapunov theorem. The Lyapunov function has the form

$$V = \frac{1}{2}\left(\tilde{i}_{rx}^2 + \tilde{i}_{ry}^2 + \tilde{H}_x^2 + \tilde{H}_y^2 + \tilde{\theta}_r^2\right) > 0, \tag{120}$$

where.

$$\tilde{i}_{rx} = \hat{i}_{rx} - i_{rx}, \tilde{i}_{ry} = \hat{i}_{ry} - i_{ry}, \tilde{H}_x = \hat{H}_x - H_x, \tilde{H}_y = \hat{H}_y - H_y \text{ and } \tilde{\theta}_r = \hat{\theta}_r - \theta_r. \tag{121}$$

The proposed observer structure will be asymptotically stable if $\dot{V}_2 < 0$ and if the stabilizing functions introduced to the structure are determined as

$$v_{rx} = -c_x \tilde{i}_{rx}, \tag{122}$$

$$v_{ry} = -c_y \tilde{i}_{ry}, \tag{123}$$

$$v_{Hx} = c_{Hx}\left(-\frac{L_s}{w_\sigma}\tilde{i}_{ry} - \hat{\omega}_r R_r \tilde{i}_{rx}\right), \tag{124}$$

$$v_{Hy} = c_{Hy}\left(\frac{L_s}{w_\sigma}\tilde{i}_{rx} + \hat{\omega}_r R_r \tilde{i}_{ry}\right), \tag{125}$$

where $(c_x, c_y, c_{Hx}, c_{Hy}, c_\theta) > 0$ are the observer tuning gains and

$$v_\theta = -c_\theta \tilde{\theta}_r. \tag{126}$$

The speed observer structure will be asymptotically stable if (122)–(126) is satisfied. In the sensorless control, the rotor speed is not measured, therefore the deviation $\tilde{\theta}_r$ in (126) should be replaced by $\tilde{\theta}_H$. This means that the deviation between the estimated values of H_x and H_y, calculated from (113)–(114) and estimated from the observer structure in (117)–(118) is as follows:

$$\tilde{\theta}_H = \tan^{-1}(\vartheta), \tag{127}$$

where

$$\vartheta = \frac{H_x\hat{H}_y - H_y\hat{H}_x}{H_x\hat{H}_x + H_y\hat{H}_y} \text{ and } \left(H_x\hat{H}_x + H_y\hat{H}_y\right) \neq 0. \tag{128}$$

The value $\hat{H}_x, \hat{H}_y$ can be estimated from

$$\hat{H}_x = \hat{\omega}_r\left(L_m i_{sx} + L_r\hat{i}_{rx}\right), \tag{129}$$

$$\hat{H}_y = \hat{\omega}_r\left(L_m i_{sy} + L_r\hat{i}_{ry}\right). \tag{130}$$

In the observer structure, the rotor speed is not estimated adaptively, therefore the rotor speed error in (120) is not considered. The rotor speed value can be estimated from the non-adaptive scheme. From (129)–(130), after some calculation, the rotor speed value can be determined from

$$\hat{\omega}_r = \frac{\hat{H}_x\hat{\psi}_{rx} + \hat{H}_y\hat{\psi}_{ry} - c_f s_\omega}{\hat{\psi}_{rx}^2 + \hat{\psi}_{ry}^2}, \tag{131}$$

where

$$s_\omega = \hat{H}_x \hat{\psi}_{ry} - \hat{H}_y \hat{\psi}_{rx}, \quad (132)$$

$$\hat{\psi}_{rx} = L_m i_{sx} + L_r \hat{i}_{rx}, \quad (133)$$

$$\hat{\psi}_{ry} = L_m i_{sy} + L_r \hat{i}_{ry}, \quad (134)$$

$c_f \geq 0$ and $\left(\hat{\psi}_{rx}^2 + \hat{\psi}_{ry}^2\right) \neq 0$.

In **Figure 10a**, the responses of DFIG on the active and reactive power changes are shown. After 0.1 s, the active power s_p value is changed from -0.1 to -0.35, and reactive power is set to -0.6 p.u. and changed at the same time. After 0.4 s, the active and reactive powers are changed s_p to 0.35 and s_q to 0.2 p.u. The rotor speed estimation error is smaller than 0.015 in the dynamic states, the same as the rotor position error.

In **Figure 10b**, the active power s_p is set to 0.02 p.u. and reactive power s_q is set to -0.6 p.u. The rotor speed of the DFIG is changed from the sub-synchronous to super-synchronous mode. Close to synchronous rotor speed (1.0 p.u.), the estimated speed error was smaller than 0.01 p.u., and it is increasing when the speed is growing. The rotor position error has almost the same value.

In **Figure 11a**, the active power is changed from -0.1 to -0.35 p.u. The rotor speed estimation error is smaller than 0.05 p.u. and the rotor position is smaller than 0.1 p.u. during these changes (in the experimental results). In **Figure 11b**, the reactive power is changed from -0.7 to -0.4 p.u., the rotor speed error is smaller than 0.035 p.u., and the rotor position error value is changed from 0.05 to -0.05 p.u.

In **Figure 12a**, the rotor speed crosses from -1.1 (super-synchronous mode) to -0.7 p.u. (sub-synchronous mode). The estimated value of rotor position is growing and close to synchronous speed (-1.0 p.u.) achieving the value of about 0.12. The reactive power value was set to -0.6 p.u. With LPF, the filtered value of stabilizing

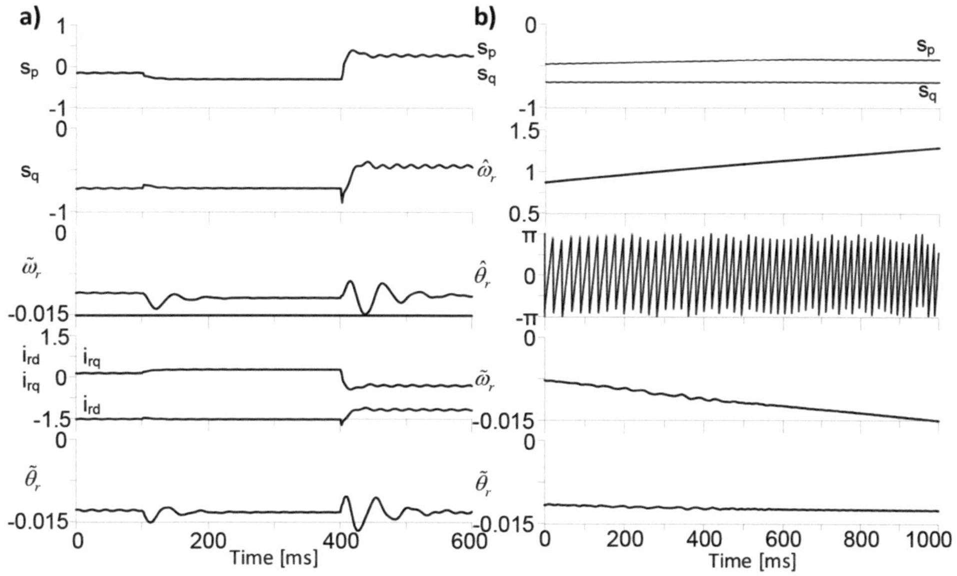

Figure 10.
a) the changes of the active and reactive power, b) sub and super-synchronous working modes – Simulation results.

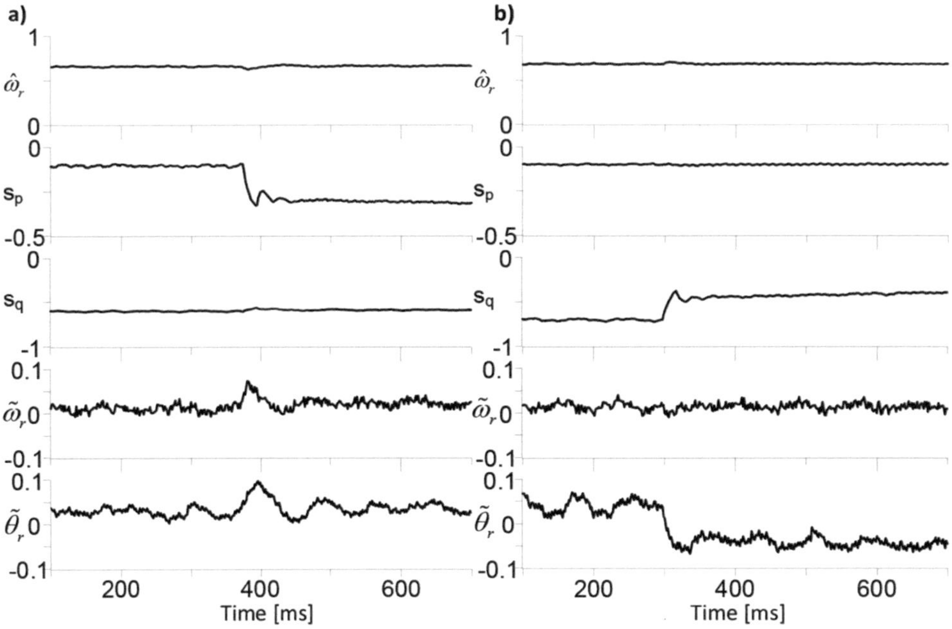

Figure 11.
The changes: a) the active power from −0.1 to −0.35 p.u., b) reactive power from −0.7 to −0.4 p.u. – Experimental results.

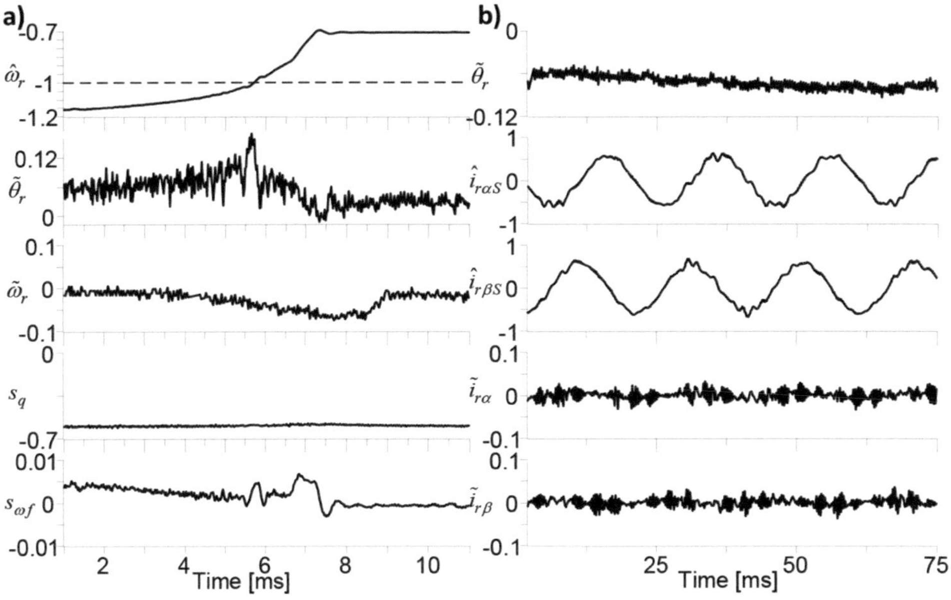

Figure 12.
The waveforms of chosen variables in a) the rotor speed are changed from the super-synchronous to sub-synchronous mode, b) the steady state – Experimental results.

function $\hat{s}_{\omega f}$ is presented. The value of this function is about 0.005 for the super-synchronous mode and about 0.001 p.u. for the sub-synchronous mode. The estimated rotor speed error is about 0.05 p.u. during the crossing through the synchronous speed.

6. Conclusions

In this chapter, robust mechanism for different structures of speed observers or rotor position was presented. The solution was tagged "robust mechanism" because of the introduction of stabilizing function in the speed or rotor position estimation schemes. The additional stabilization law prevents the unstable working range of the speed observer structure (positive poles of the observer). In this chapter, the stability analysis based on the Lyapunov function was presented. The introduced additional stabilizing function to the observer structure is based on Lagrange's identity, which is the main contribution of this chapter. The form of the proposed robust mechanism is based mainly on the vector and scalar product of the two chosen vectors in the observer system. The mutual position of these vectors directly influences the position of the estimated vectors of the observer and also influences the estimated rotor speed value or the rotor position. The mutual position of a vector influences the value of the estimated electromagnetic torque of the machine. The proposed solution has significant meaning during the low speed of the IM and IPMSM (due to the unstable working points of the observer structure), as well as during the synchronous rotor speed of the DFIG system. The proposed robust mechanism for the speed estimation scheme can be applied to each observer structure, which is based on the space vector form of the mathematical model of an observer system.

Nomenclature

"$\wedge$"	estimated values,
"$\sim$"	error of estimated values,
$i_{sx,y}$	stator current vector components,
$i_{rx,y}$	stator current vector components,
$u_{r\alpha,\beta}$	rotor voltage vector components,
$u_{s\alpha,\beta}$	stator voltage vector components,
ω_r	rotor angular speed,
θ_r	rotor position,
R_r, R_s	rotor and stator resistances,
L_m	mutual-flux inductance,
L_s, L_r	stator and rotor inductances,
T_e	electromagnetic torque,
T_L	load torque,
J	machine torque of inertia,
τ	relative time,
$\hat{\theta}_r$	estimated rotor position,
$\hat{\omega}_r$	estimated rotor electrical speed,
$\tilde{\omega}_r$	rotor speed error,
$\tilde{\theta}_r$	rotor position error,
(x-y)	coordinate system associated with any angular speed,
(d-q)	coordinate system associated with the rotor angular speed,
L_d, L_q	stator winding inductances,
ψ_f	permanent magnet flux linkage,
IM	induction machine,

IPMSM	interior permanent magnet synchronous machine,
DFIG	doubly fed induction generator,
AFO	adaptive full-order observer.

Author details

Marcin Morawiec
Department of Electric Drives and Energy Conversion, Gdańsk University of Technology, Gdańsk, Poland

*Address all correspondence to: marcin.morawiec@pg.edu.pl

References

[1] Orlowska-Kowalska T, Korzonek M, Tarchała G. Stability analysis of selected speed estimators for induction motor drive in regenerating mode—A comparative study. IEEE Transactions on Industrial Electronics. 2017;**62**(10)

[2] Kubota H, Matsuse K, Nakano T. "DSP-based speed adaptive flux observer of induction motor". IEEE Transactions on Industry Applications. Mar./Apr. 1993;**29**:344-348

[3] Yin Z, Li G, Zhang Y, Liu J. Symmetric-strong-tracking-extended-Kalman-filter-based Sensorless control of IM drives for Modeling error reduction. IEEE Transactions on Industrial Informatics. 2019;**15**:650-662

[4] Comanescu M. Design and implementation of a highly robust sensorless sliding mode observer for the flux magnitude of the induction motor. IEEE Transactions on Energy Conversion. 2016;**2**:649-657

[5] Morawiec M. Z type observer backstepping for induction machines. IEEE Transactions on Industrial Electronics. 2015;**62**(4):2090-2103

[6] Morawiec M, Kroplewski P, Odeh C. Nonadaptive rotor speed estimation of induction machine in an adaptive full-order observer. IEEE Transactions on Industrial Electronics. 2022;**69**:2333-2344

[7] Kubota H, Matsuse K. Speed Sensorless field-oriented control of induction motor with rotor resistance adaptation. IEEE Transactions on Industry Applications. 1994;**30**: 1219-1224

[8] Sunwankawin S, Sangwongwanich S. Design strategy of an adaptive full-order observer for speed-sensorless induction motor drive – Tracking performance and stabilization. IEEE Transactions on Industrial Electronics. 2006;**53**(1): 96-119

[9] Sun W, Liu K, Jiang D, Ou R. Zero synchronous speed stable operation strategy for speed Sensorless induction motor drive with virtual voltage injection. IEEE Energy Conversion Congress and Exposition (ECCE). 2018; **2018**:337-343. DOI: 10.1109/ECCE.2018.8557756

[10] Chen J, Huang J. Globally stable speed-adaptive observer with auxiliary states for Sensorless induction motor drives. IEEE Transactions on Power Electronics. 2019;**34**(1):33-39

[11] Hinkkanen M, Luomi J. "Stabilization of regenerating-mode operation in sensorless induction motor drives by full-order flux observer design". IEEE Transactions on Industrial Electronics. Dec. 2004;**51**(6): 1318-1328

[12] Glumineau A, Morales JL. Sensorless AC Electric Motor Control Robust Advanced Design Techniques and Applications. Cham: Springer; 2015

[13] Gou L, Wang C, You X, Zhou M, Dong S. IPMSM sensorless control for zero- and low-speed regions under low switching frequency condition based on fundamental model. IEEE Transactions on Industrial Electronics. 2021;**2021**

[14] Kim S, Im J, Song E, Kim R. A new rotor position estimation method of IPMSM using all-pass filter on high-frequency rotating voltage signal injection. IEEE Transactions on Industrial Electronics. 2016;**63**: 6499-6509

[15] Ioannou P, Sun J. Robust Adaptive Control. Englewood cliffs, NJ: Prentice Hall; 1996

[16] Howard A, Rorres C. Relationships between dot and cross products. In: Elementary Linear Algebra: Applications Version. 10th ed. Hoboken, NJ: John Wiley & Sons; 2010. p. 162

[17] Laskhmikanthan V, Leela S, Martynyuk A. Practical Stability of Nonlinear Systems. Singapore: Word Scientific; 1990

[18] Traoré D, Plestan F, Glumineau A, Leon J. Sensorless induction motor: High-order sliding-mode controller and adaptive interconnected observers. IEEE Transactions on Industrial Electronics. 2008;**55**(11):3818-3827

[19] Krzeminski Z. A new speed observer for control system of induction motor. In: IEEE Int. Conference on Power Electronics and Drive Systems, PESC'99. Hong Kong; 1999

[20] Xiaoyuan W, Pingxin W, Jiahong Y, Gengji W. Study the Effects of Slotted Rotor on q-axis Direction on Permanent Magnet Synchronous Motor. In: 15th International Conference on Electrical Machines and Systems (ICEMS). Sapporo, Japan: Hokkaido Citizens Actives Center; 2012

[21] Jung S, Kobayashi H, Doki S, Okuma S. An improvement of sensorless control performance by a mathematical modelling method of spatial harmonics for a SynRM. In: International Power Electronics Conference - ECCE ASIA. 2010. pp. 2010-2015

[22] Agrawal J, Bodkhe S. Experimental study of low speed Sensorless control of PMSM drive using high frequency signal injection. Power Engineering and Electrical Engineering. 2016;**14**(1). DOI: 10.15598/aeee.v14i1.1564

[23] Yang S, Ajjarapu V. A speed-adaptive reduced-order observer for Sensorless vector control of doubly fed induction generator-based variable-speed wind turbines. IEEE Transactions on Energy Conversion. 2010;**25**(3): 891-900. DOI: 10.1109/TEC.2009.2032589

[24] Morawiec M, Blecharz K, Lewicki A. Sensorless rotor position estimation of doubly-fed induction generator based on backstepping technique. IEEE Transactions On Industrial Electronics. 2020;**67**(7):5889-5899

Chapter 5

Classical Direct Torque Control for Switched Reluctance Motor Drive

V. Pushparajesh and B.M. Nandish

Abstract

The modern electrical machines require higher efficiency in concern with pollution of the environment. Industries are focusing on bringing out new avenues in controlling the electric motors to adjust the speed and torque without compromising. The Direct Torque Control technique is suggested in this study. Slip control, which exploits a peculiar link between slip and torque, is the basic concept underlying this regulation. Direct torque control provides various benefits over field-oriented control, including reduced sensitivity to machine parameters, easy assembly, and quick dynamic torque response. As the voltage space vector is chosen in response to the inaccuracy in the flux linkage and torque, a current controller is unnecessary in this design. Low torque ripple, reduced noise, and reduced mechanical vibration are all attainable through proper torque management in the switching reluctance motor.

Keywords: direct torque control (DTC), field-oriented control, dynamic torque response, flux linkage, slip control, torque control

1. Introduction

Because of its simple mechanical structure, low cost, efficiency, The Switched Reluctance Motor (SRM) has the potential to become one of the most widely used low-cost electromechanical energy converters due to its advantageous torque/speed characteristic and very minimal requirement for maintenance. However, this drive's non-uniform torque output characteristics and doubly salient construction mean it generates more noise and torque ripples, limiting its usefulness. As a result, various methods have been developed for reducing torque ripple in switching reluctance motors. Several torque control techniques are studied with the goal of enhancing the drive's efficiency through reduced torque ripple and faster response times.

The early 1980s saw the development of Direct Torque Control for use with AC drives. In 2012, Yong Chang Zhang, et al., proposed a new direct torque control for three-level inverter supplied AC drives [1]. By adjusting the state voltage vectors in relation to the torque and flux errors, we are able to exert direct control over the torque. Direct torque control for a switching reluctance motor was developed by Moron et al. using the lyapunov function [2]. In order to precisely control the torque applied to a switching reluctance motor, Sahoo et al. presented a lyapunov function [3]. It was proposed by Yong Chang Zhang et al. in 2012 that a sensorless drive for a three-level inverter-fed induction motor could benefit from enhanced direct torque control [1]. A low-ripple torque control at high speeds was implemented by Jin Ye et al. for switching reluctance motor drives [4].

Improvements in switching reluctance motor performance were studied by Qingguo Sun et al., who looked into the role of direct torque control and torque sharing function [5]. Conventional direct torque control of the four-phase switching reluctance motor was created by Srinivas Pratapgiri and Prasad Polaki Venkata Narsimha in 2012 [6]. As the bandwidth of the hysteresis controller is restricted in this control, the decrease of torque ripple is minimal at best. For switching reluctance motors, proposed a method of shared control of current and flux linkage [7]. For switching reluctance motors, Jipun and Luk achieved sensorless direct torque control [8]. This article discusses a machine with a shorter flux path and modifies the electrical and mechanical phases so that they both add up to 45 degrees. When it comes to the direct torque management of a switching reluctance motor, Bosra et al. proposed a four-level converter [9].

Due to the odd number of phases, there is a paucity of material on direct torque control with four phase SRM. Compared to field orientation control, the DTC's many benefits include reduced reliance on machine parameters, a quicker dynamic torque response, and a more straightforward design. Any current controller is unnecessary for the DTC as the voltage space vectors are chosen in response to flux linkage and torque faults. Direct torque control, or DTC, is a method whereby a motor's torque and speed are adjusted in response to changes in the motor's electromagnetic field.

2. Control strategy

When the converter switches, the motor's flux and torque are directly controlled, making direct torque control an optimal AC variable frequency control concept. **Figure 1** depicts the rudimentary DTC block diagram. Based on the estimated flux and torque and the reference flux and torque, the stator voltage and resistance may be calculated. Based on the output levels of this hysteresis comparator, the error is sent to the hysteresis comparator, and the switching table is used to determine the voltage vector that will be supplied to the voltage source inverter in order to obtain the reference torque. Direct torque control for a four-phase switching reluctance motor is

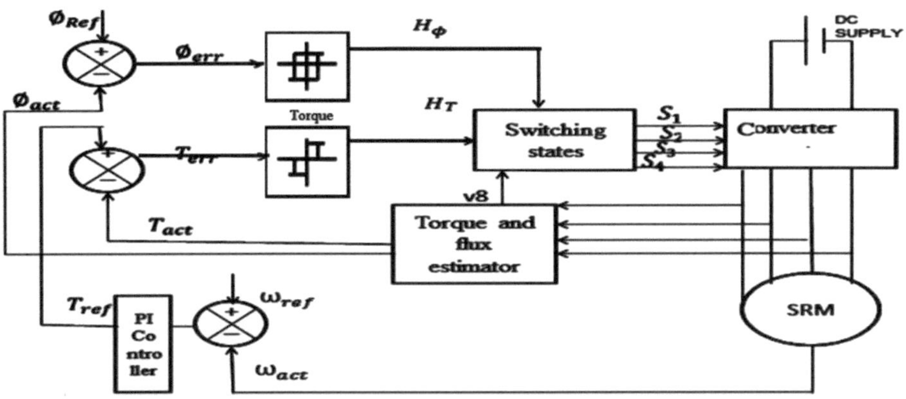

Figure 1.
Block diagram of classical DTC.

outlined, together with its underlying principles, the specific steps involved, and the means for putting it into practise [10].

Because of their unique four-phase, eight-by-six-polar configuration, synchronous motors cannot be controlled using the same direct torque technique as inductance motors. SRM employs the reluctance principle for producing torque, with each phase functioning separately and sequentially. Torque is generated in either a positive or negative direction depending on the magnitude of the change in stator flux amplitude in relation to the rotor's location. We call the former "flux acceleration" and the latter "flux deceleration," respectively, when the value is positive or negative. Therefore, the following constitutes a definition of a novel approach to SRM regulation [11].

- The stator flux linkage vector of the motor is kept within amplitude hysteresis bands.
- During the stator flux vector acceleration or deceleration, the torque can be controlled.

The control goal is accomplished by varying the voltage vector and speeding up or slowing down the stator flux vector in relation to the rotor rotation [12]. The magnitude of the torque is also a function of the instantaneous current, which is different from the way things are handled with traditional control. It is also shown that the stator current in this drive control system exhibits a first order delay with respect to the variation in stator flux. In this way, it is safe to assume that the current remains stable even if the flux is sped up or slowed down under control [13]. This permits the control method to regulate torque solely with respect to the value of the flux acceleration and deceleration, and independently of the current change. This is similar to the traditional control scheme, which assumes that the rotor flux remains unchanged despite variations in the stator flux and modifies the motor's torque via regulation of the stator flux acceleration [14].

3. Impact of voltage vectors

The Direct Torque Control loop has a torque hysteresis controller with three levels (1, 0, and −1) and a flux hysteresis controller with two levels (1 & −1). The SR motor has a distinctive pole structure, hence the voltage space vector for each phase is said to be perpendicular to the pole's central axis. Keep in mind that the physical winding topology of the motor has not been altered from its standard setting.

Each motor phase can be in one of three voltage states, determined on the drive's circuit topology configuration. A zero-voltage loop arises and is defined as the condition "0" when current is flowing with one device off (V1 or V2). Similar to how the condition is described as "−1" when both devices (V1&V2) are switched off and "1" when both devices (V1&V2) are turned on, the motor phase experiences a negative voltage when both are off and a positive value when both are on. **Table 1** displays the three voltage states that can exist in a four-phase SRM drive, which is a result of the switching function.

As a result, there are (mn, where m represents no of voltage states and n represents no of phases) 81 distinct permutations, as opposed to just two for the classical DTC of the AC motor. **Figure 2** depicts the eight alternative spatial locations of switching voltage vectors for defining the voltage states, which are defined in the same way as in the standard direct torque control algorithm but with equal amplitude voltage vectors that are separated by radians. The mathematical model of IM has typically been

Switching stage of power converter	Terminal voltage of winding	Switching function S
V1 and V2 both on	Positive Voltage	1
VI and V2 one On and the other Off	Zero Voltage	0
Vl and V2 both Off	Negative voltage	−1

Table 1.
Switching function table.

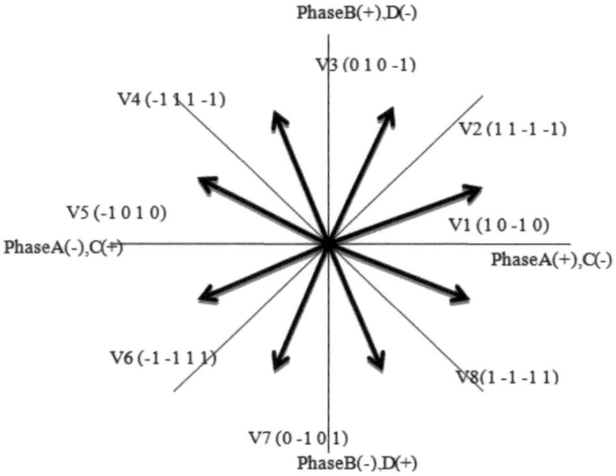

Figure 2.
Spatial location of switching voltage vectors.

analysed using the dq coordinate system. Within a two-axis rotating reference frame, the abc to dq0 transformation determines the direct axis, quadratic axis, and zero sequence qualities of a three-phase sinusoidal signal. The park transformation is a standard tool for modelling three-phase electric machines. Because the stator and rotor values can be referred to a stationary or rotating reference frame, time-varying inductances can be eliminated.

Each stator winding's flux linkage is assumed to be at the magnetic pole's centreline for the sake of convenience. **Figure 3** shows the stable α-β coordinate as a result. In Eqs. (1) to (4), and Ψ_α and Ψ_β stand for the two-flux linkage that flows in the two equivalent rotors, generating the same flux as the stator ψ_1, ψ_2, ψ_3 and ψ_4 currents (1.4).

$$\Psi\alpha = \frac{1}{\sqrt{2}}\left[\psi1 + \psi2\ \cos 45 - \psi4\ \cos 45\right] \tag{1}$$

$$\Psi\beta = \frac{1}{\sqrt{2}}\left[\psi3 + \psi2\ \sin 45 + \psi4\ \sin 45\right] \tag{2}$$

$$\Psi s = \sqrt{\Psi\alpha^2 + \Psi\beta^2} \tag{3}$$

$$\delta = \text{arct}\left(\Psi_{\beta/}\ \Psi_\alpha\right) \tag{4}$$

where, Ψ_s is instantaneous composition flux linkage, is spatial position angle of composition flux linkage.

Hysteresis controller		Voltage vector selection							
Φ	T	S (1)	S (2)	S (3)	S (4)	S (5)	S (6)	S (7)	S (8)
1	1	U_2	U_3	U_4	U_5	U_6	U_7	U_8	U_1
	0	U_0	U_9	U_0	U_9	U_0	U_9	U_0	U_9
	−1	U_7	U_8	U_1	U_2	U_3	U_4	U_5	U_6
0	1	U_3	U_4	U_5	U_6	U_7	U_8	U_1	U_2
	0	U_0	U_9	U_0	U_9	U_0	U_9	U_0	U_9
	−1	U_6	U_7	U_8	U_1	U_2	U_3	U_4	U_5

Table 2.
Switching voltage vector for converter.

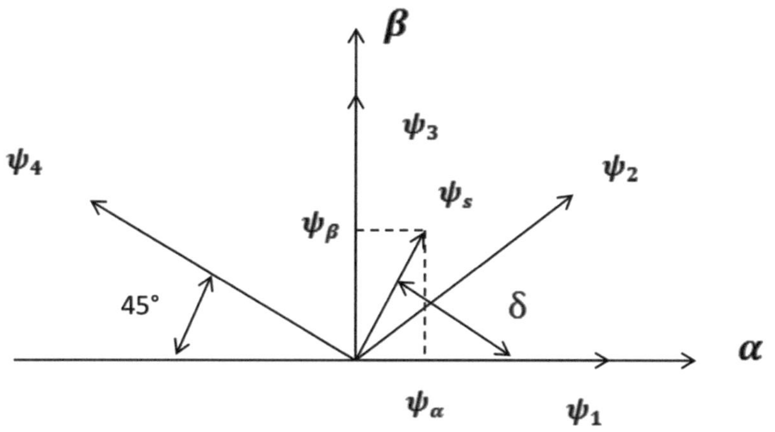

Figure 3.
Composition flux linkage vector of switched reluctance motor.

Coordinate decomposition concept yields composition flow linkage height that is 1.4% greater than energy conservation approach, as seen in the aforementioned formulae. As a result, that SRM will spend a lot of time operating at the magnetic saturation point. While steady error is still ensured, motor efficiency drops dramatically towards magnetic saturation. Using the α-β vector block, we can determine which part of the plane the flow vector occupies. Sector of the plane in which the flow vector lies is one of eight, with each sector separated from the others by 45 degrees. If the stator flux linkage is in the kth region, increasing the flux with the switching vectors U k+1 and U k−1, or decreasing it with U k+3 and U k−3, is possible. Switching vectors U k +1 and U k +3 can be used to enhance the torque, whereas U k−1 and U k−3 can be used to decrease it. **Table 2** illustrates the converter's voltage switching vector selection. There are eight "active" voltage vectors labelled U1 through U8 and two "null" vectors labelled U0 and U9.

3.1 Classical direct torque control results

The aforementioned technique is used to model the direct torque control scheme implemented in a MATLAB/Simulink simulation of a four-phase switching reluctance motor drive. One torque hysteresis controller with three levels and another flux hysteresis controller with two levels are used to create the simulation model. With reference to equations 1.2–1.4, the, α, β flux transformation is performed. The phase count and rotor-position angle inform the design of the flux sector. The voltage switching vectors form the basis for the lookup table. In the virtual experiment, a 4-phase motor is used. Each zone's switching vector is chosen from the vector table based on the output signals from the two-hysteresis comparator, which are in turn determined by the position sensor. The stator flux hysteresis band in this control approach is set to 0.01Wb, and the torque hysteresis band is set to 0.3Nm; both values are held within these bounds throughout all simulations. Within 50 milliseconds, the motor will have reached the desired speed, having drawn a starting current that was capped by the converter's components' maximum ratings. The tests are performed at a constant stator flux of 0.3Wb under a wide range of load torques.

Using a proportional integral controller with kp and ki set to 0.10 and 0.01 respectively, the speed error is transformed into the reference torque. The simulation was run to examine how well the hysteresis controller worked under varying speeds and loads. The 1.5 kilowatt, 3800 revolution per minute, 4 Newton-meter reluctance drive is put through its paces with a dc voltage of 120 volts. It has been decided that 10mH will be the aligned Inductance and 49mH will be the unaligned Inductance. We've decided on a value of 0.008kg.m.m for the moment of inertia and 0.01N.m.s for the coefficient of friction. Below are examples of switching reluctance motor drive performance under a variety of load and speed scenarios.

4. Performance at rated load condition

When first activated, the switching reluctance drive is subjected to the rated load torque and speed. In **Figure 4**, we see the current and torque response in this situation. The phase current and total torque time scale variation is assumed to be 0.71–0.76 seconds for simplicity in interpreting the responses. It can be seen from the curve that the rated condition yields a maximum current of 8A.

The hysteresis controller minimises torque inaccuracy by choosing the best switching vector, which in turn minimises torque ripple as depicted in the image. Using the following Equation, one can get both the total torque production and the torque ripple in percentage terms (5).

$$T_{ripple} = \left[(T_{max} - T_{min})/T_{avg}\right]^{*} 100\% \tag{5}$$

The figure is further enlarged to specify the variation of the torque ripple accurately at rated torque over a wide range of speed and is shown in **Figure 5**.

The response curve (a) indicates that the torque achieves a minimum of 3.94 Nm and a maximum of 4.15 Nm, with an average of 4.0 Nm reached at 0.09 s. Torque is 5.2% of the total estimated force. According to the torque response curve(b), the specified torque is reached in 0.09 seconds, after which the output torque ranges between a maximum of 4.18 Nm and a minimum of 3.92Nm. An average torque value of 4.01Nm was measured. The percentage of torque ripple is 6.47 percent. In the torque response curve (c), the rated torque is reached in 0.09 s, following which the output torque ranges between a maximum of 4.2 Nm and a minimum of 3.9 Nm. Torque production (T_{avg}) is measured to be 4.05 Nm on average. Torque ripple is 7.4 percent, according to the calculations.

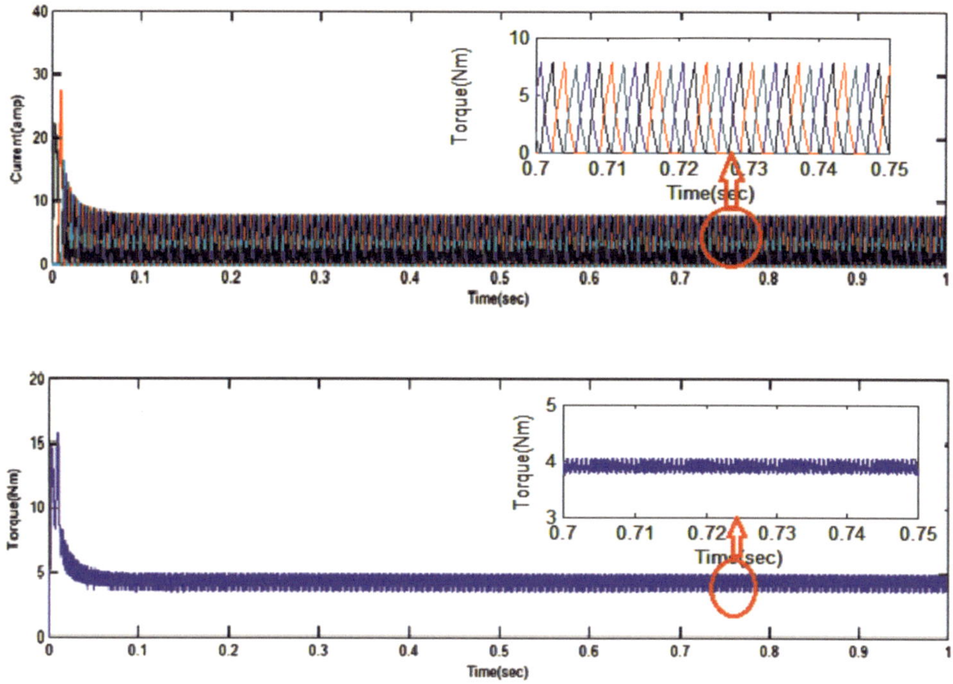

Figure 4.
Current and torque response of the motor at rated load condition.

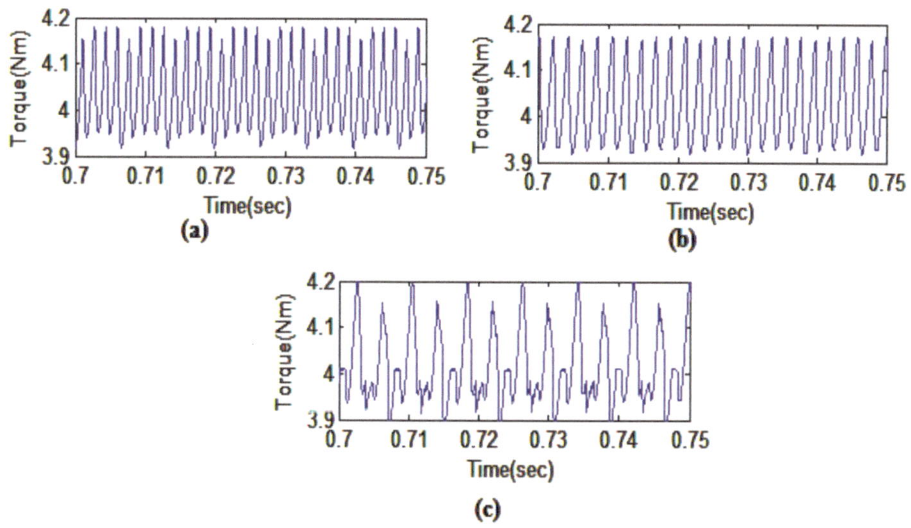

Figure 5.
Torque ripple waveforms at rated torque (a) Rated speed (b) Half rated speed (c) 10% of rated speed.

5. Performance at 75% of the rated load

At first, the SRM drive is stimulated at 75% load torque at a variety of rated speeds. **Figure 6** displays the resulting current and torque response. According to the graph, the rated condition produces a maximum current of 6A.

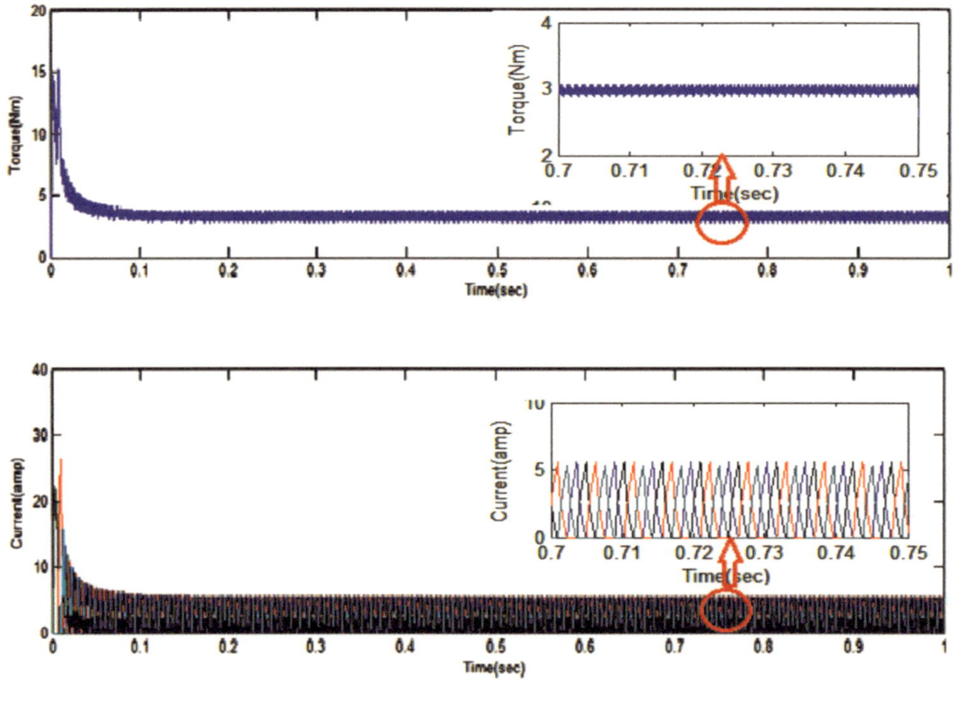

Figure 6.
Phase current and torque response at 75% of the rated load with rated speed.

The suggested controller outperforms the alternative proposed controller in terms of torque response and current response under the aforementioned conditions. **Figure 7** enlarges the torque response curve displayed in **Figure 6** to show the precise fluctuation of torque under 75% of the rated load situation with respect to the extensive range of speed variation.

In 0.08 seconds, the output torque (T_{out}) reaches 75% of the specified torque, and from there it varies between a maximum of 3.13 Nm and a minimum of 2.98 Nm, as

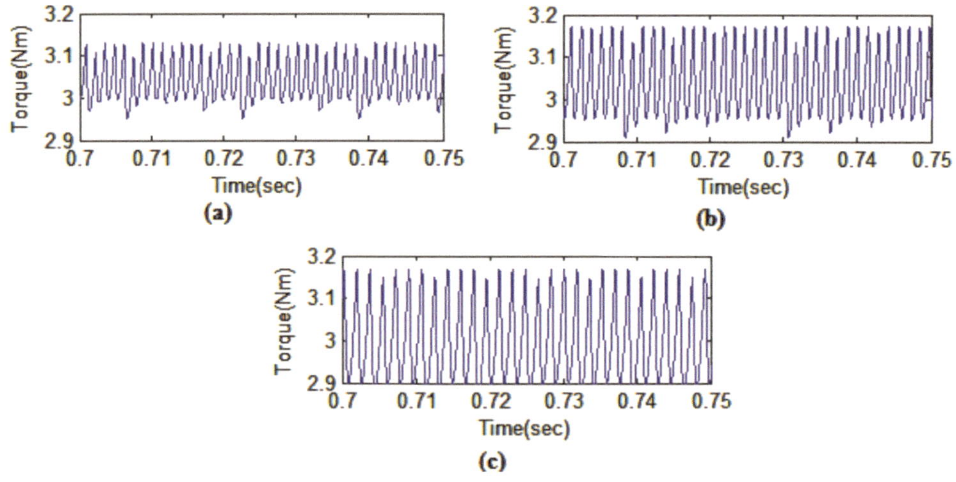

Figure 7.
Torque response at 75% of the rated load (a) Rated speed (b) half of rated speed (c) 10% of rated speed.

shown in the response curve (a). Generally speaking, 3.0Nm is the average torque production (T_{avg}) that was measured. Torque ripple is 5.0% as determined by the calculation.

Torque output (T_{out}) reaches 75% of rated torque in 0.08 seconds, as shown by the response curve (b), and then ranges between a maximum of 3.18 Nm and a minimum of 2.98 Nm. Torque production (T_{avg}) is measured to be 3.1 Nm on average. 6.4% is the percentage value derived for the torque ripple. Torque output (T_{out}) reaches 75% rated torque in 0.08 seconds, and after that, it ranges between a maximum of 3.12 Nm and a minimum of 2.90 Nm, as shown by the torque response curve (c). It has been determined that the average torque output (T_{avg}) is 3.05 Nm, and that the torque ripple is 7.2% of that value. The output torque (T_{out}) achieves 75% of rated torque in 0.08 s, as shown by the response curve (a), and then ranges between a maximum (T_{max}) of 3.13 Nm and a minimum (T_{min}) of 2.98 Nm.

Generally speaking, 3.0Nm is the average torque production (T_{avg}) that was measured. Torque ripple is 5.0% as determined by the calculation. Torque output (T_{out}) reaches 75% of rated torque in 0.08 seconds, as shown by the response curve (b), and then ranges between a maximum of 3.18 Nm and a minimum of 2.98 Nm. Torque production (T_{avg}) is measured to be 3.1 Nm on average. 6.4% is the percentage value derived for the torque ripple. Torque output (T_{out}) reaches 75% rated torque in 0.08 seconds, and after that, it ranges between a maximum of 3.12 Nm and a minimum of 2.90 Nm, as shown by the torque response curve (c). It has been determined that the average torque output (T_{avg}) is 3.05 Nm, and that the torque ripple is 7.2% of that value (**Figure 8**).

In order to clearly demonstrate the fluctuation in torque under 50% of the rated load condition with respect to the different speed conditions, the above depicted torque response curve has been extended. **Figure 9** demonstrates the observable range of values.

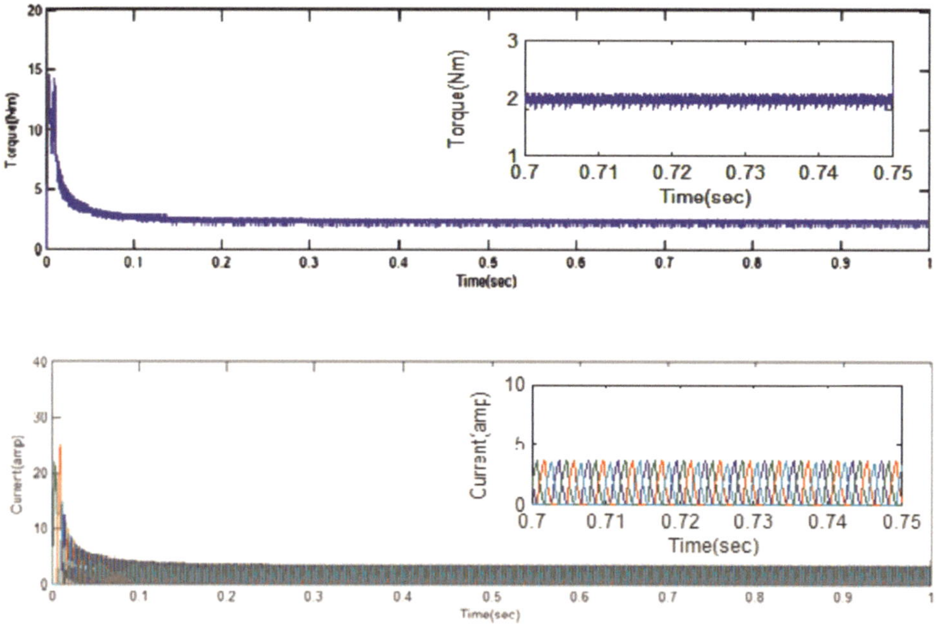

Figure 8.
Current and torque response at 50% of the load torque with rated speed.

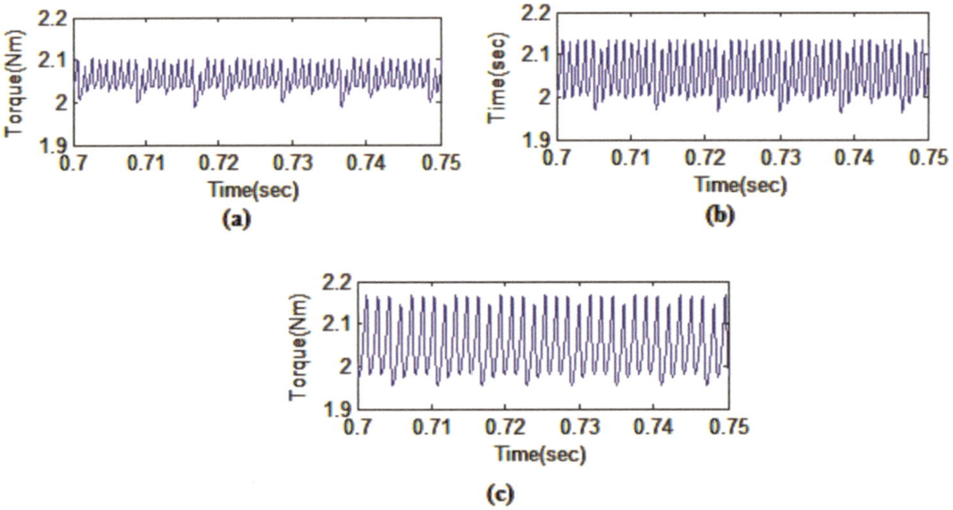

Figure 9.
Torque response at 50% of the rated load (a) Rated speed (b) half rated speed (c) 10% of rated speed.

In 0.08 seconds, the output torque (T_{out}) reaches 50% of its rated value, and from there it varies between a maximum (T_{max}) of 2.1 Nm and a minimum (T_{min}) of 2.0 Nm, as shown in the response curve (a). Torque production (T_{avg}) is calculated to be 2.02 Nm on average. 4.9% is the percentage value determined to be the torque ripple output. The torque output (T_{out}) reaches 50% of the specified torque in 0.08 s, and then it ranges between 2.11 Nm and 1.99 Nm (T_{max} and T_{min}, respectively) as shown in the torque response curve (b). Torque production (T_{avg}) was measured to be 2.0Nm on average. 6 percent is the calculated ripple percentage of torque. The torque output (T_{out}) reaches 50% of the specified torque in 0.08 seconds, as shown by the curve (c), and then ranges between a maximum (T_{max}) of 2.13 Nm and a minimum (T_{min}) of 1.99 Nm.

Generally speaking, we can say that the torque output (T_{avg}) is 2.02 Nm. Torque ripple is determined to be 6.9% of total output.

6. Performance at 25% of the rated load torque

In order to trigger the SRM dive, a 25% load torque is applied while the speed is varied. We evaluate the performance of the proposed controller by measuring and tabulating the torque and current responses. **Figure 10** depicts this rated-speed torque and current response. This curve shows that at the rated condition, the maximum current is 2A.

The aforementioned torque response curve is simplified to clearly demonstrate the accurate fluctuation of torque under 25% of the rated load situation with respect to the broad range of speed variation. **Figure 11** displays this variance.

It can be seen in the torque output profile (a) that the output torque (T_{out}) reaches 25% of the specified torque in 0.07 seconds, and then ranges between a maximum of 1.01 Nm and a minimum of 0.99Nm. The computed proportion of torque ripple output is 2.0%, and the average torque output (T_{avg}) is 1.00Nm.

Torque output (T_{out}) achieves 25% rated torque in 0.07 s, as shown by the torque response curve (b), and then ranges between a maximum of 1.015 Nm and a minimum of 0.99 Nm. To be precise, T_{avg} is 1.0Nm, which is the average torque production. According to the numbers, the torque ripple is 2.5%.

DOI: http://dx.doi.org/10.5772/intechopen.107876

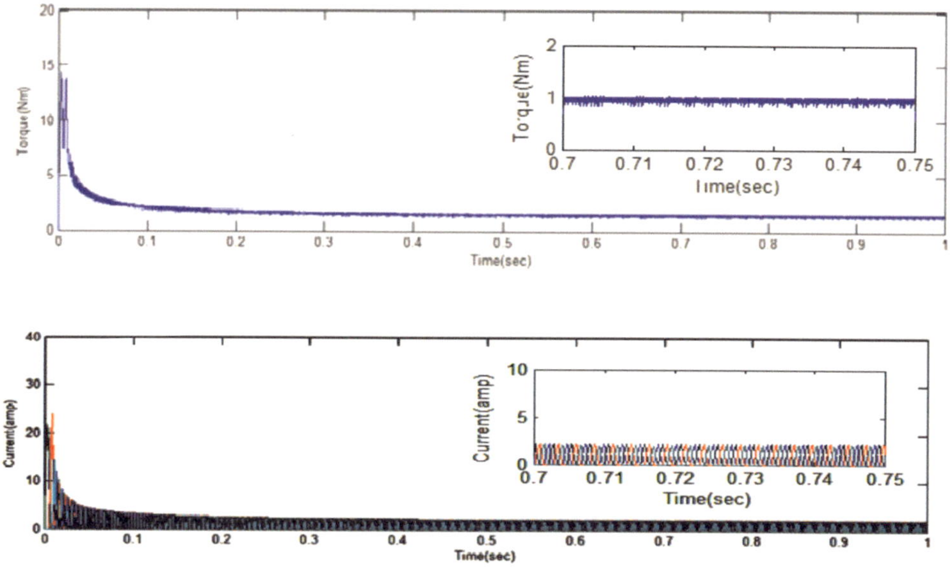

Figure 10.
Current and torque profile for 25% of the load torque at rated speed.

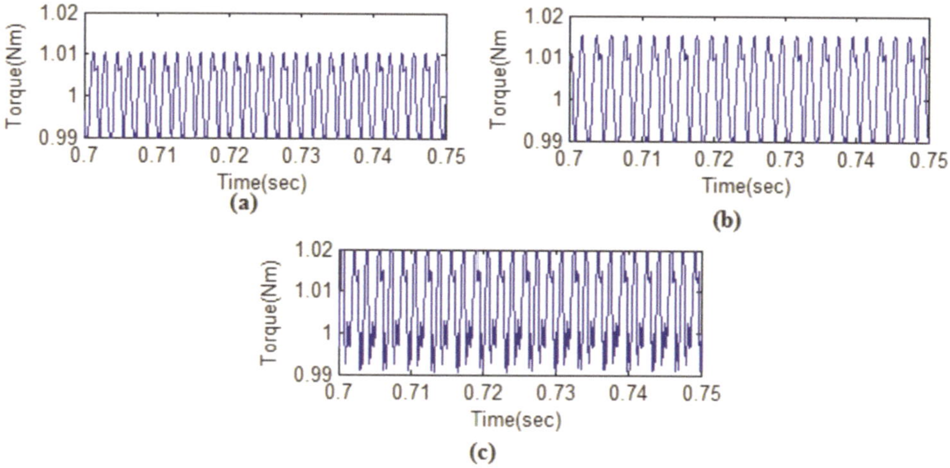

Figure 11.
Torque profile for 25% of the load torque (a) Rated speed (b) half rated speed (c) 10% of rated speed.

Torque output (T_{out}) achieves 25% rated torque in 0.07 s, as suggested by the response curve (c), and then ranges between a maximum of 1.02 Nm and a minimum of 0.99 Nm. Torque production (T_{avg}) averaged out to be 1.01 Nm. The percentage of torque output is 3.0%.

7. Performance under variable load condition

In the dynamic simulation, the speed was maintained at a constant rated condition throughout. There were two potential scenarios: It took 0.34 seconds to double the

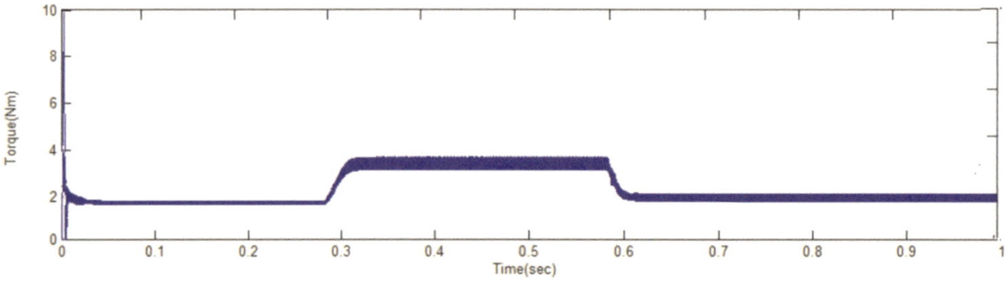

Figure 12.
Torque profile under external load variation.

command torque T_{com} from the first case shown in **Figure 12** to the second case whereas it took 0.65 seconds to halve the torque back down to 2 N.m from 4 N.m. According to the figure, the controller has excellent dynamic performance, with T_{out} increasing to 2 Nm in just 0.06s.

The most significant drawback of the conventional direct torque control is its slow response to initial torque and flux changes. Changes between steady state and step-up states use the same vectors, making it impossible to differentiate between large and tiny errors in flux and torque. Both AI and conventional techniques of control over AI can alleviate these issues. To smooth out the driving torque at any rated speed and torque, the next few chapters will focus on such smart controllers.

8. Comparative study

An analysis is made on Neural Network Controller and the revealed observations are made; **Table 3** shows the performance comparisons between different controllers.

Sl. no	Proposed controller based DTC	Rated speed in %	Torque ripple in % with respect to the % applied load torque				Computational time (sec)
			100	75	50	25	
1	Hysteresis controller	100	6.0	5.0	4.9	2.0	Equal to run time
		50	6.48	6.4	6	2.5	
		10	7.4	7.2	6.9	3.0	
2	Fuzzy controller	100	4.7	4.0	3.5	1.5	4.34
		50	5.7	5.6	5.4	2.1	
		10	6.6	6.2	5.9	2.4	
3	Neural network controller	100	3.5	3.3	2.5	1.3	3.17
		50	4.4	3.98	3.5	1.8	
		10	5.6	5.2	5.0	2.2	

Table 3.
Performance comparison between different controllers.

9. Conclusion

The direct torque control technique is tested for switched reluctance motor, Direct torque Control technique is able to minimize the ripple content in the motor torque output at different operating conditions. the torque ripple is almost minimized in the range of about 1.5% to 2% for a fixed speed with variable torque. The settling time of the torque and the response time of the speed is also reduced, which in turn increases the efficiency of the machine. The major drawback with the proposed controller is the fixation of weights during the real time application which reduce the flexibility and adaptability of the system. This drawback or limitation can be overcome by using hybrid intelligent controller.

Author details

V. Pushparajesh* and B.M. Nandish
Department of Electrical and Electronics, FET, Jain Deemed to be University, Karnataka, India

*Address all correspondence to: pushparajesh.v@gmail.com

References

[1] Zhang YC, Zhu J, Zhao Z, Xu W, Dorrell DG. An improved direct torque control for three-level inverter-fed induction motor sensorless drive. IEEE Transactions on Power Electronics. 2012; **27**(3):1502-1513

[2] Moron C, Garcia A, Tremps E, Somolinos JA. A lyapunov function based robust direct torque controller for a switched reluctance motor drive system. IEEE Transactions on Magnetics. 2012;**48**(4):1661-1664

[3] Sahoo SK, Dasgupta S, Panda SK, Xu JX. A Lyapunov function-based robust direct torque controller for a switched reluctance motor drive system. IEEE Transactions on Power Electronics. 2012;**27**(2):555-564

[4] Ye J, Bilgin B, Emadi A. An extended-speed low-ripple torque control of switched reluctance motor drives. IEEE Transactions on Power Electronics. 2015; **30**(3):1457-1470

[5] Sun Q, Wu J, Gan C, Shen M. Investigation of direct torque control and torque sharing function strategy for switched reluctance motor applications'. In: Proceedings of the 18th International Conference on Electrical Machines and Systems. Pattaya, Thailand. 2015. pp. 864-868

[6] Pratapgiri S, Polaki P, Narsimha V. Direct torque control of 4 phase 8/6 switched reluctance motor drive for constant torque load. World Journal of Modelling and Simulation. 2012;**8**(3): 185-195

[7] Wang J-J. Direct torque control of 4 phase 8/6 switched reluctance motor drive for constant torque load. Electrical Power System Research. 2016; **131**:19-30

[8] Jipun P, Luk PC. Direct torque control of Sensor less switched reluctance motor drives. In: Proceedings of the Seventh International Conference on Power Electronics and Variable Speed Drives. London, UK. 1998. pp. 329-334

[9] Bosra MM, Saniei M, Kianinezhad R. Direct torque control of switched reluctance motor using four level convertor. In: Proceedings of the Nineteenth Irarian Conference on Electrical Engineering. Tehran, Iran. 2011. pp. 1-6

[10] Min Y, Shengjie J, Xiaogang Y. Direct torque control for four phase switched reluctance motor. In: 2010 International Conference on Computer, Mechatronics, Control and Electronic Engineering. Changchun, China. 2010. pp. 531-534

[11] Pushparajesh V, Nandish BM, Marulasiddappa HB. Hybrid intelligent controller based torque ripple minimization in switched reluctance motor drive. Bulletin of Electrical Engineering and Informatics;**10**(3): 1193-1203

[12] Pushparajesh V, Nandish BM, Marulasiddappa HB. Torque ripple minimization in switched reluctance motor using ANFIS controller. In: 2021, WSEAS Transactions on Systems and Control. 2021. pp. 171-182

[13] Cheok AD, Fukuda Y. A new torque and flux control method for switched reluctance motor drives. IEEE Transactions on Power Electronics;**17**:543-557

[14] Mohamed N, Hamza T, Brahim G. Novel DTC induction machine drive improvement using controlled rectifier for DC voltage tuning. International Journal of Power Electronics and Drive Systems (IJPEDS). 2021;**10**:1223-1228